Flutter
从0到1构建大前端应用

何瑞君◎著

电子工业出版社
Publishing House of Electronics Industry
北京·BEIJING

内 容 简 介

本书是 Flutter 从基础入门到进阶实战的教程书，也是一本面向大前端的新技术书。Flutter 是谷歌新推出的一个跨平台的、开源的 UI 框架，同时支持 iOS 系统和 Android 系统开发，并且是谷歌未来新操作系统 Fuchsia 的默认开发套件。本书共 10 章，内容包括 Flutter 简介、Dart 语言入门、一切皆组件、事件处理、动画、使用网络技术与异步编程、路由、持久化、插件与混合工程和项目实战。本书不仅介绍了 Flutter 的基本原理、特性，而且在实战章节全面展示了打造一个完整的基于 Flutter 的 App 的流程，包含具体细节、思想流程和代码实操。

本书适合 Flutter 初学者，有一定移动开发（iOS/Android）经验的人员，以及希望了解 Flutter 原理并进阶实战的相关技术人员。

未经许可，不得以任何方式复制或抄袭本书之部分或全部内容。
版权所有，侵权必究。

图书在版编目（CIP）数据

Flutter：从 0 到 1 构建大前端应用 / 何瑞君著. --北京：电子工业出版社，2019.7
ISBN 978-7-121-36179-1

Ⅰ. ①F… Ⅱ. ①何… Ⅲ. ①移动终端－应用程序－程序设计 Ⅳ. ①TN929.53

中国版本图书馆 CIP 数据核字（2019）第 053415 号

策划编辑：陈　林
责任编辑：张彦红
印　　刷：北京季蜂印刷有限公司
装　　订：北京季蜂印刷有限公司
出版发行：电子工业出版社
　　　　　北京市海淀区万寿路 173 信箱　邮编：100036
开　　本：720×1000　1/16　印张：20　字数：288 千字
版　　次：2019 年 7 月第 1 版
印　　次：2019 年 10 月第 3 次印刷
定　　价：79.00 元

凡所购买电子工业出版社图书有缺损问题，请向购买书店调换。若书店售缺，请与本社发行部联系，联系及邮购电话：（010）88254888，88258888。
质量投诉请发邮件至 zlts@phei.com.cn，盗版侵权举报请发邮件至 dbqq@phei.com.cn。
本书咨询联系方式：（010）51260888-819，faq@phei.com.cn。

前　　言

开发者的思考

随着移动开发技术的发展与成熟，移动端所处的萌芽阶段也早已结束。现在，iOS 与 Android 系统已经越来越成熟，各种 App 之间的系统差异性也越来越小，可以说，移动开发技术已经进入了"下半场"。其中，如何构建大前端的跨平台方案，是近年来十分火热的话题之一。

我在工作之余，喜欢浏览一些技术类网站，比如掘金、CSDN 等。在这些技术网站上，我了解和学习了不少新技术，也在实践这几年出现的跨平台技术，比如 Hybrid、React Native、Weex 等。

在 2018 年年初，我第一次了解到谷歌的 Flutter 技术。经过一番实践之后，我认为这是一种很有前景的新技术。它可以快速在 iOS 和 Android 系统上构建高质量的原生用户界面。很明显，Flutter 为大前端的跨平台方案提供了一个全新的思路。为此，我专门做了一些开源项目并发表了一些文章，以便更深入地学习、推广 Flutter 技术。非常荣幸，在这个过程中，电子工业出版社的陈林老师找到了我，并希望我写一本关于 Flutter 的技术书。这着实让我受宠若惊，也倍感责任重大。

Flutter 是一门新技术，大家都处于学习的状态中。为了便于大家理解，也便于自己的技术积累，我在学习过程中不断总结、做笔记，逐渐整理和创作了《Flutter：从 0 到 1 构建大前端应用》一书。

读者对象

这是一本关于 Flutter 入门的书，从基础讲起，也会涉及与 Dart 语言相关的内容，以便于读者快速迈向 Flutter 开发。同时，本书也会涉及技术原理、思维等拔高内容。

本书通俗易懂，由浅入深，既适合初学者学习，也适合专业人员阅读。对于在 Android、iOS、前端等领域有过相关工作经验的读者来说，阅读体验会更好。

本书内容

本书内容是非常系统化的，用 10 章讲述了学习 Flutter 必须掌握的知识，涉及 Flutter 简介、环境搭建、Dart 语言简介、组件、事件处理、动画、网络、路由、持久化、插件和实战项目等。实践项目会教你如何构建一个 Flutter 应用并使用 Node.js 把服务端搭建起来。此外，还会专门写一个 Flutter 异常上报的项目，用于错误日志的跟踪。

本书各章内容比较独立，你可以按照顺序阅读，进行从 0 到 1 的全面学习；也可以根据需要把本书作为一本工具查询书，直接跳转到需要查询的章节。各章的内容摘要如下所示。

第1章，Flutter简介：从整体上先介绍移动端近年的发展变化，然后引出Flutter，并介绍Flutter的环境搭建。

第2章，Dart语言入门：介绍要编写Flutter项目所必须掌握的Dart语言核心语法知识，为学好Flutter做铺垫。

第3章，一切皆组件：介绍Flutter相关的核心组件和使用场景等。

第4章，事件处理：介绍Flutter的事件处理机制等。

第5章，动画：介绍Flutter的动画相关内容与核心原理等。

第6章，使用网络技术与异步编程：介绍Flutter网络技术的相关内容、网络层与服务端的交互，以及Flutter的异步编程等。

第7章，路由：介绍Flutter的路由跳转方式和原理等。

第8章，持久化：介绍Flutter的几种持久化存储方式。

第9章，插件与混合工程：介绍Flutter的插件编写方式以及如何在现有原生项目里加入Flutter的相关技术。

第10章，项目实战：第一个项目从服务端与客户端的角度，介绍一个完整的项目案例；第二个项目介绍Flutter的日志捕获方式以及服务端采集日志的方式。

源码下载

建议你在学习本书的时候先把相关的案例源码运行一遍，然后按照自己的思路写一写，这样可以加深印象。切勿直接复制、粘贴、运行源码，只有亲手实践才会有所收获。本书源码在Flutter 1.5.4版本中是兼容的，读者可以正常使用和练习。

源码下载地址：https://github.com/heruijun/FlutterFrom0To1。

技术交流

这本书虽然是关于新技术的，但也代表了我多年技术经验的积累和技术思维的沉淀。当然，书中所写内容难免会有纰漏之处，希望读者朋友能及时指正，希望我们相互学习、共同进步。

读者可以加入 Flutter 技术交流 QQ 群（群号：468010872），群里有不少技术"大牛"和前端专家，也可以在群内反馈和交流本书学习心得。

目　　录

第 1 章　Flutter 简介 ·· 1

1.1　Flutter 的优势 ·· 1
1.2　对比其他技术 ·· 2
1.3　Flutter 架构 ··· 3
　　1.3.1　Flutter Framework ·· 3
　　1.3.2　Flutter Engine ·· 4
1.4　开发环境搭建 ·· 4
　　1.4.1　Mac 上的环境搭建 ··· 4
　　1.4.2　在 Mac 上获取 SDK 并设置环境变量 ······································· 5
　　1.4.3　安装 Xcode 与运行模拟器 ·· 6
　　1.4.4　安装 Android Studio ··· 7
　　1.4.5　在 Android Studio 上安装 Flutter 开发插件 ····························· 8
　　1.4.6　安装 VSCode 与 Flutter 开发插件 ·· 9
　　1.4.7　IDE 的选择 ·· 11
　　1.4.8　使用 Flutter 诊断工具检查 Flutter 开发环境 ·························· 11
　　1.4.9　创建 Demo 工程并体验热重载 ··· 12
1.5　Flutter 升级 ··· 14
本章小结 ·· 15

第 2 章 Dart 语言入门 ·· 16

2.1 应用场景 ·· 16
2.1.1 SDK 安装和升级 ·· 17
2.1.2 编写一个 HelloWorld 并运行 ·· 17

2.2 变量与常量 ·· 18
2.2.1 变量 ·· 18
2.2.2 常量 ·· 18
2.2.3 内置类型 ·· 19
2.2.4 数值型 ·· 19
2.2.5 数值型操作 ·· 20
2.2.6 字符串 ·· 21
2.2.7 字符串操作 ·· 21
2.2.8 布尔型 ·· 23
2.2.9 List 与数组 ·· 23
2.2.10 Map ·· 24
2.2.11 dynamic 和 Object ·· 25

2.3 运算符 ·· 26

2.4 异常捕获 ·· 27

2.5 函数 Function ·· 28
2.5.1 main 函数 ·· 28
2.5.2 可选参数 ·· 28
2.5.3 必传参数 ·· 29
2.5.4 可选的位置参数 ·· 29
2.5.5 默认参数 ·· 29
2.5.6 函数作为参数传递 ·· 30
2.5.7 函数作为变量 ·· 30

2.6 异步编程 ·· 30
2.6.1 Future 是什么 ·· 31
2.6.2 async 和 await ·· 31
2.6.3 继承、接口实现和混合 ·· 33
2.6.4 泛型 ·· 37

本章小结 ·· 39

第 3 章 一切皆组件 ··· 40

3.1 基础组件（Basic widgets） ··· 42
3.1.1 Text ··· 43
3.1.2 Icon ··· 43
3.1.3 Image ·· 45
3.1.4 Button ·· 45
3.1.5 FlutterLogo ··· 47

3.2 单一子元素组件（Single-child） ·· 48
3.2.1 Container ··· 48
3.2.2 Container 的约束 ··· 49
3.2.3 SingleChildScrollView ··· 52
3.2.4 FittedBox ··· 53
3.2.5 FractionallySizedBox ·· 54
3.2.6 ConstrainedBox ·· 55
3.2.7 Baseline ··· 56
3.2.8 IntrinsicWidth 和 IntrinsicHeight ····························· 58

3.3 多子元素组件（Multi-child） ·· 58
3.3.1 Scaffold ··· 58
3.3.2 AppBar ·· 59
3.3.3 Row 和 Column ·· 61
3.3.4 ListView ·· 64
3.3.5 GridView ··· 69
3.3.6 CustomScrollView ··· 70
3.3.7 Flex ·· 73
3.3.8 Wrap ··· 75

3.4 状态管理 ·· 77
3.4.1 Widget 树 ·· 78
3.4.2 Context ·· 79
3.4.3 StatelessWidget ··· 80
3.4.4 StatefulWidget ·· 80
3.4.5 StatefulWidget 的组成 ··· 81

          3.4.6 State ··· 82
          3.4.7 State 生命周期 ·· 82
          3.4.8 Widget 的唯一身份标识：key ·· 88
          3.4.9 InheritedWidget ·· 89
    3.5 包管理 ··· 93
    3.6 常用代码段效果 ··· 94
          3.6.1 案例一：侧滑效果 ·· 95
          3.6.2 案例二：登录界面 ·· 96
          3.6.3 案例三：轮播图效果 ·· 96
          3.6.4 案例四：图片浏览器的相册效果 ··· 97
          3.6.5 案例五：全局主题设置 ··· 97
    本章小结 ··· 98

第 4 章 事件处理 ·· 99
    4.1 原始指针事件 ··· 99
          4.1.1 基本用法 ··· 99
          4.1.2 忽略事件 ··· 102
    4.2 GestureDetector ·· 105
          4.2.1 基本用法 ··· 105
          4.2.2 常用事件 ··· 106
          4.2.3 拖曳和缩放效果 ·· 109
          4.2.4 事件竞争与手势冲突 ·· 112
          4.2.5 手势识别器 ·· 114
    4.3 事件原理与分发机制 ·· 117
    4.4 事件通知 ··· 122
    本章小结 ··· 125

第 5 章 动画 ··· 126
    5.1 动画原理及概述 ·· 126
          5.1.1 Animation ·· 127
          5.1.2 Animatable ·· 127
          5.1.3 AnimationController ·· 128

	5.1.4	Tween	130
	5.1.5	Tween.animate	133
	5.1.6	Curve	133
5.2	动画的封装与简化		136
	5.2.1	AnimatedWidget	136
	5.2.2	AnimatedBuilder	137
5.3	Hero 动画		141
	5.3.1	基本用法	141
	5.3.2	实现原理	143
5.4	交错动画		145
5.5	动画示例		151
	5.5.1	自定义加载动画	151
	5.5.2	实现动画效果	153
	5.5.3	Dialog 加载框	157
	5.5.4	测试加载框效果	159
本章小结			161

第 6 章 使用网络技术与异步编程 162

6.1	网络协议简介		162
	6.1.1	HTTP 协议简介	163
	6.1.2	HTTP 2.0 能给我们带来什么	164
	6.1.3	HTTPS	166
6.2	网络编程		167
	6.2.1	HttpClient	167
	6.2.2	http 库	170
6.3	JSON 解析		171
	6.3.1	JSON 转成 Dart 对象	172
	6.3.2	一个完整的例子	173
	6.3.3	根据 JSON 用工具生成实体类	175
6.4	dio 库		176
	6.4.1	基本用法	177
	6.4.2	dio 单例	177

	6.4.3 dio 拦截器	178
	6.4.4 dio 拦截器链	180
	6.4.5 dio 适配器	181
	6.4.6 dio 库总结	183
6.5	异步编程	184
	6.5.1 isolate	184
	6.5.2 event loop	184
	6.5.3 线程模型与 isolate	188
	6.5.4 创建单独的 isolate	190
	6.5.5 Stream 事件流	192
本章小结		195

第 7 章 路由 196

7.1	路由简介	196
	7.1.1 基本用法	197
	7.1.2 静态路由	197
	7.1.3 动态路由	200
	7.1.4 参数回传	203
7.2	路由栈	206
	7.2.1 路由栈详解	207
	7.2.2 pushReplacementNamed 方法	207
	7.2.3 popAndPushNamed 方法	208
	7.2.4 pushNamedAndRemoveUntil 方法	209
	7.2.5 popUntil 方法	210
7.3	自定义路由	210
本章小结		214

第 8 章 持久化 215

8.1	shared_preferences 本地存储	215
	8.1.1 shared_preferences 的常用操作	216
	8.1.2 shared_preferences 举例	216

8.2 SQLite 数据库 ······ 219
8.2.1 sqflite 依赖库简介 ······ 220
8.2.2 封装 SQL Helpers ······ 223
8.2.3 sqflite 实现员工打卡示例 ······ 225
8.3 文件形式存储 ······ 232
8.3.1 path_provider 简介 ······ 233
8.3.2 一个简单的日记本示例 ······ 233
本章小结 ······ 236

第 9 章 插件与混合工程 ······ 237
9.1 package ······ 237
9.1.1 添加 package 的几种方式 ······ 238
9.1.2 更新 package ······ 239
9.1.3 创建自己的 package ······ 240
9.1.4 发布 package ······ 240
9.2 理解 Platform Channel ······ 241
9.2.1 消息传递与编解码器 ······ 242
9.2.2 Platform 数据类型支持 ······ 243
9.2.3 MethodChannel 简介 ······ 243
9.2.4 SharedPreferences 插件源码解析 ······ 245
9.3 混合开发 ······ 247
9.3.1 创建 Flutter 模块 ······ 248
9.3.2 关联原生工程 ······ 248
9.3.3 编写混合工程代码 ······ 249
9.3.4 热重载混合端代码 ······ 251
9.3.5 aar 模块化打包 ······ 252
9.4 FlutterBoost 混合方案 ······ 253
9.4.1 框架的由来 ······ 253
9.4.2 使用 FlutterBoost 改进 ······ 254
9.4.3 FlutterBoost 源码分析 ······ 255
本章小结 ······ 263

第 10 章 项目实战 ··· 264

10.1 实战一：实现一个招聘类 App ··· 264
10.1.1 项目需求与技术选型 ··· 264
10.1.2 服务端设计 ··· 265
10.1.3 Flutter 基础架构 ··· 270
10.1.4 启动页面 ··· 271
10.1.5 使用 dio 实现网络请求 ··· 273
10.1.6 公司列表与详情实现 ··· 275
10.1.7 用 WebSocket 实现聊天模块 ··· 281

10.2 实战二：实现异常上报系统 ··· 286
10.2.1 实现原理 ··· 286
10.2.2 FlutterError.onError 和 Zone ··· 287
10.2.3 异常上报 Flutter 的实现 ··· 288
10.2.4 异常上报 Android 端的实现 ··· 290
10.2.5 服务端接收异常上报 ··· 293

10.3 实战项目源码 ··· 299
10.4 性能分析与辅助工具 ··· 300
本章小结 ··· 304

第 1 章

Flutter 简介

　　Flutter 是谷歌新推出的一套跨平台的、开源的 UI 框架，同时支持 iOS、Android 系统开发，并且是未来新操作系统 Fuchsia 的默认开发套件。Flutter 自 2017 年 5 月发布第一个版本以来，更新迭代了近 60 个版本，并且在 2018 年 12 月初发布了 1.0 稳定版。笔者成书时的版本已经是 1.5.4 版了。从 Flutter 团队开发的那么多版本再结合当前的发展形势来看，谷歌正在大力推广 Flutter。在 Stack Overflow 上，关于 Flutter 的提问越来越多，而在 GitHub 上与此相关的社区活跃度也不断高涨。下面，我们先概括性地介绍 Flutter 的一些基本情况。

1.1　Flutter的优势

　　综合来看，Flutter 的优势主要有以下几方面。

　　（1）跨平台性。真正做到一套代码可以同时用在 Android 和 iOS 两大平台，避免过高的维护成本，节省测试、开发资源。

　　（2）通过"自绘 UI+原生系统"实现高帧率的流畅 UI。不使用 WebView 这种比较老的开发模式，而使用 Skia 作为 2D 渲染引擎，使用 Dart 语言作为运行时，以及使用 Text 作为文字排版的引擎。

　　（3）支持开发过程中的热重载。在开发过程中，只需要保存操作步骤就可以刷新 Flutter 项目，提高开发效率。

（4）对开发环境要求不高，支持 Android Studio 和 Visual Studio Code（VSCode），更为轻量级。

（5）高性能。直接调用系统的 API 绘制 UI，因此，性能更接近原生。

（6）学习成本低。如果之前学习过 React Native（RN），那么再学习 Flutter 也会很快就能掌握。如果具有前端或者纯原生开发经验，则学起来也比较省力。

1.2　对比其他技术

在 Flutter 诞生之前，有许多其他的客户端技术，比如 Hybrid、RN、Weex。这些技术都广泛运用在各大 App 中。通过表 1.1 我们看一下常用的客户端技术之间有什么不同。

表 1.1

技术类型	UI 渲染方式	性能	开发效率	动态化	取值
HTML5+原生	WebView 渲染	一般	高	支持	Cordova、Ionic
JavaScript+原生渲染	原生控件渲染	好	高	支持	RN、Weex
自绘 UI+原生	调用系统 API 渲染	好	Flutter 高，Qt 低	默认不支持	Qt、Flutter

从表 1.1 来看，实现自绘 UI 的还有 Qt，但其开发效率比 Flutter 低。本书就不对此做扩展介绍了，有兴趣的读者可以查阅相关资料。

其中 RN 和 Weex 都是通过 JavaScript 与原生系统交互，并通过 JavaScript 发送布局消息传递给原生端，实现布局渲染的。而 HTML5+原生（Hybrid 技术）则通过 WebView 容器实现，并且由 JSBridge 与原生系统直接通信。Flutter 的 Release 包默认是使用 Dart AOT 模式编译的，所以不支持动态化（动态化指的是动态下发送代码实现热更新）。不过在 Dart 语言中还有 JIT 和 SnapShot 这类运行方式，这些模式都是支持动态化的。

1.3 Flutter架构

我们学习一门新技术，首先要对其整个技术体系有一个大概的了解。脑海里先要有一些相关的框架和轮廓，就好比造房子，要有施工图一样，写代码也要先有架构图，然后依据核心架构图，再去实现技术上的需求。我们来看一张官方的架构图，如图 1.1 所示。

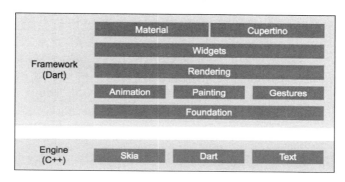

图 1.1

图 1.1 形象地展示了 Flutter 的架构图，它分为两部分，即 Framework 和 Engine。

1.3.1 Flutter Framework

Framework 是由纯 Dart 语言实现的 SDK（软件开发工具包）。Dart 是 Flutter 的官方语言，第 2 章我们将重点介绍 Dart 语言的内容。根据图 1.1，介绍一下相关层的具体内容和作用。

（1）底下两层：底层 UI 库，提供动画、手势及绘制能力，是谷歌暴露给开发者调用的。

（2）Rendering 层：它负责构建 UI 树，每当 UI 树上的 Element 发生变化时，都会计算出有变化的部分并且会更新 UI 树，最终将 UI 树绘制到屏幕上展示给用户。

（3）Widgets 层。基础组件库，Flutter 默认帮我们提供了 Material 和 Cupertino 两种视觉风格的组件库。在大多数情况下，官方提供的 UI 库可以满足我们日常的开发。

1.3.2　Flutter Engine

Engine 是由纯 C++语言实现的 SDK，根据字面意思，我们了解到 Engine 是引擎的意思。Framework 层所有的 UI 库调用都会用到 Engine 这一层。

（1）Skia：2D 渲染引擎（Android 系统自带，但 iOS 系统不自带，因此，iOS 包所占的存储空间更大）。

（2）Dart：Dart 运行时。

（3）Text：文字排版引擎。

1.4　开发环境搭建

官方已经发布了 1.5.4 稳定版，基于该版本，我们以图文的形式来介绍一下 Flutter 开发环境是怎样搭建的以及开发插件是如何安装的。如果读者对安装环境搭建已经比较了解，可以跳过本节。

1.4.1　Mac 上的环境搭建

官方列出了 Mac 需要的系统配置，具体如下所示。

（1）操作系统：macOS（64-bit）。

（2）剩余磁盘空间：700 MB（不包括集成开发环境和工具）。

（3）命令行工具：Mac 一般都自带，Flutter 依赖下面这些命令行工具。

- bash

- curl

- git 2.x

- mkdir

- rm

- unzip

- which

1.4.2　在 Mac 上获取 SDK 并设置环境变量

可以在官方网址查看最新的 SDK 版本，地址是 https://flutter.io/docs/get-started/install/macos，具体步骤如下所示。

（1）默认会列出最新版本的下载地址，如图 1.2 所示。

flutter_macos_v1.5.4-stable.zip

图 1.2

（2）下载完之后，解压缩。可以通过 unzip 命令解压缩，如下所示：

```
cd ~/development
unzip ~/Downloads/flutter_macos_v1.5.4-stable.zip
```

（3）设置环境变量，找到之前解压存放的 Flutter SDK 路径，然后在当前用户的.bash_profile（没有则新建一个）里指向 Flutter SDK 目录，内容如下所示：

```
export PATH="$PATH:[PATH_TO_FLUTTER_GIT_DIRECTORY]/flutter/bin"
```

注意：把[PATH_TO_FLUTTER_GIT_DIRECTORY]替换为 SDK 存放目录。

（4）设置完之后，需要刷新一下".bash_profile"，输入以下命令：

```
source $HOME/.bash_profile
```

刷新完之后，可以输入"echo $PATH"来验证是否设置成功。

注意：如果不是第一次安装 Flutter，在已经安装过的情况下，可以通过"flutter upgrade"命令对 Flutter SDK 进行升级。

1.4.3　安装 Xcode 与运行模拟器

Xcode 是为 iOS 开发者提供的开发者工具。由于 Flutter 是跨平台的，因此在 macOS 环境下需要安装 Xcode，安装步骤如下所示。

（1）需要先安装 Xcode 9.0 或更高的版本（可以通过官方下载地址或 App Store 安装）。

（2）配置 Xcode 命令行工具以确保使用最新安装的 Xcode 版本。在一般情况下，当你想要使用最新版本的 Xcode 时，正确的路径是 sudo xcode/select/switch/Applications/Xcode.app/Contents/Developer。如果你需要使用不同的版本，比如还安装了 Xcode 的一些 Beta 版本，就需要指定相应的路径。

（3）在命令行中运行"sudo xcodebuild-license"并同意 Xcode 条款。

（4）输入命令"open -a Simulator"来打开模拟器，如图 1.3 所示。

图 1.3

在学习了创建 Flutter 项目之后，进入项目目录，输入"flutter run"就可以让项目在模拟器中运行起来。

1.4.4　安装 Android Studio

Android Studio 是为 Android 开发者提供的 IDE，是 Flutter 的主要开发工具，安装步骤如下所示。

（1）去 Android 官方网站下载 Android Studio，下载地址是 https://developer.android.com/studio，在笔者成书时，最新版本是 3.3。

（2）在下载完之后，只需要根据 Android Studio 的安装引导过程设置 Android SDK、Platform-Tools、Build-Tools，一直点击"Next"即可安装成功。

（3）创建 Android 模拟器。可以启动"Android Studio—Tools—Android—AVD Manager"并选择"Create Virtual Device"，然后选择一个需要的设备，如图 1.4 所示。

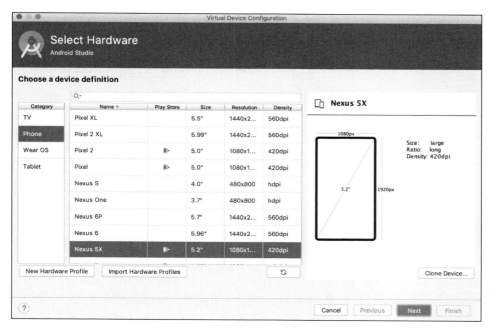

图 1.4

一直点击"Next",添加所需要的模拟器即可。

1.4.5 在 Android Studio 上安装 Flutter 开发插件

插件安装也很简单,和其他的 Android Studio 插件的安装方式完全一样。在这里,Flutter 开发环境需要安装 Flutter 插件和 Dart 插件,操作步骤如下所示。

(1)在 1.4.4 节中,我们已经完成了 Android Studio 的安装,启动 Andriod Studio。

(2)在 Andriod Studio 的"Preferences—Plugins"选项里搜索 Flutter 并安装,如图 1.5 所示。

图 1.5

(3)安装完之后,我们重启 Android Studio,发现一个新的选项"Start a new Flutter project"。如果看见这个选项,证明 Android Studio 上的 Flutter 开发插件已经安装成功,如图 1.6 所示。

图 1.6

1.4.6　安装 VSCode 与 Flutter 开发插件

VSCode 是微软出的一款 IDE，在笔者写前端 JavaScript 时会经常用到，其 IDE 界面也很酷。VSCode 打开效果如图 1.7 所示。

图 1.7

光有 IDE 是不够的，得支持 Flutter 才行。同理，和 Android Studio 一样，需要额外安装 Flutter 与 Dart 插件，步骤如下所示。

（1）启动 VSCode。

（2）在 View 菜单里选择"Command Palette"，会在屏幕上方出现一个下拉菜单。

（3）输入"install"，并选择"Extensions: Install Extension"。

（4）搜索框输入"flutter"，安装完成并重启。

（5）通过图形化方式安装也是可以的。选择 VSCode 左侧边栏的方形的"插件"按钮，然后同样可以搜索"flutter"并安装 flutter 插件，如图 1.8 所示。

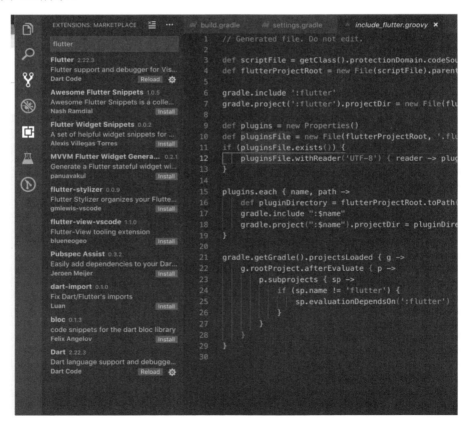

图 1.8

（6）成功安装插件后，我们可以调出"Command Palette"，然后可以调出"Flutter: New Project"。

1.4.7　IDE 的选择

在前文中，我们分别介绍了 Android Studio、VSCode 的安装和开发插件的配置。这两款 IDE 是 Flutter 官方推荐的。对于开发者来说，怎样选择，可以根据自己的喜好决定。这也体现了 Flutter 对开发环境要求不高的优势。笔者建议，如果之前是从事 Android 开发的，可以选择 Android Studio，如果之前是从事前端开发的，可以选择 VSCode。笔者个人更喜欢用 Android Studio 开发 Flutter，因此本书后续章节除了第 2 章 Dart 语言介绍会用 VSCode 开发，其他章节的代码都是基于 Android Studio 的。

1.4.8　使用 Flutter 诊断工具检查 Flutter 开发环境

在上述 Flutter SDK 和开发 IDE 安装完成之后，我们可以通过命令"flutter doctor"对 Flutter 的整体环境进行检测。如果检测下来环境都没有问题，效果会如图 1.9 所示。

```
heruijundeMacBook-Pro:~ heruijun$ flutter doctor
Doctor summary (to see all details, run flutter doctor -v):
[✓] Flutter (Channel beta, v1.1.8, on Mac OS X 10.14.2 18C54, locale zh-Hans-CN)
[✓] Android toolchain - develop for Android devices (Android SDK version 28.0.3)
[✓] iOS toolchain - develop for iOS devices (Xcode 10.1)
[✓] Android Studio (version 3.3)
[✓] VS Code (version 1.30.2)
[!] Connected device
    ! No devices available
```

图 1.9

如果有任何问题，"flutter doctor"诊断工具会给出相应的问题列表，并给出修复这些问题的具体方案。一般来说，按照本书所写几个步骤操作下来，就不会有错误提示。

1.4.9　创建 Demo 工程并体验热重载

在 1.4.8 节中，如果"flutter doctor"诊断工具检测下来运行环境没有任何问题，我们就来创建一个 Demo 工程，看一看 Flutter 为我们做了什么。

创建工程的方式前面已经介绍了，可以在 Android Studio 和 VSCode 的 "Command Palette"里创建，这里就不重复了。我们介绍一种纯命令行方式创建的方法。在终端里我们输入以下命令：

```
flutter create myapp
```

输入完之后，我们会有一个短暂的等待过程，那是 Flutter 在为我们创建一个名为"myapp"的 Flutter 项目，并且，创建完成之后，会自动执行 "flutter doctor"诊断工具来检测项目有没有问题。创建过程如图 1.10 所示。

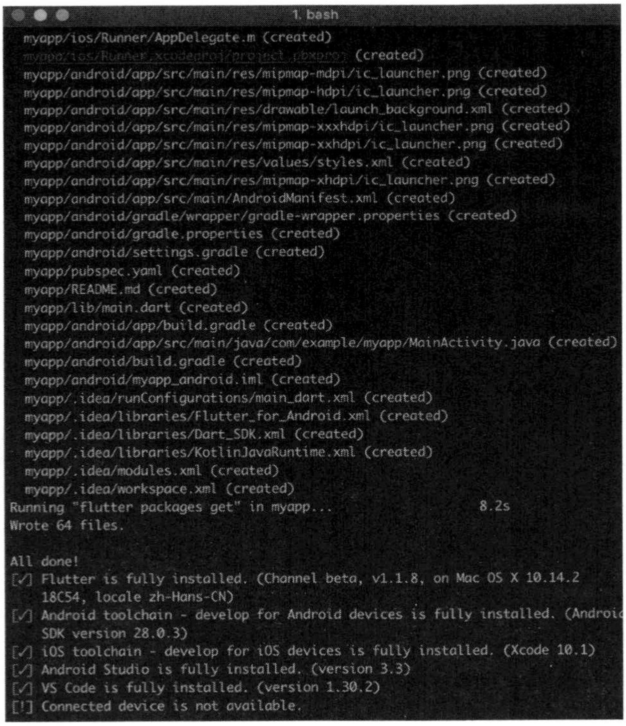

图 1.10

创建完成之后，我们可以通过 IDE 导入创建的项目。以 Android Studio 为例，打开之后选择"Open an exiting Android Studio project"，并选择前面创建的工程文件夹"myapp"，经过短暂的等待就导入完成了。不管用何种方式创建的 Flutter 工程，目录结构都是一致的，在 Android Studio 里目录层级如图 1.11 所示。

图 1.11

我们可以看到，在 myapp 里有 lib 文件夹，里面对应的是 main.dart 文件。这是官方为我们生成的入口文件，里面有一些代码和注释。我们先不关心这些代码的具体细节，只让代码运行起来，看一看效果并体验一下。

在 Android Studio 中选择模拟器（可选 iOS/Android 模拟器或真机），点击绿色的三角形的"开始"按钮运行，运行结果如图 1.12 所示。

官方的例子很简单，实现了一个简单的点击计数器。点击右下角的"加号"按钮，计数器会根据点击的次数往上增加。

接下来，我们体验一下 Flutter 的"热重载"（Hot Reload）特性。我们试着改变中间的一段文字。打开 lib/main.dart 文件，把里面的"You have pushed the button this many times:"改为"当前按钮点击次数："，改完之后直接保存 lib/main.dar 文件。这个时候，模拟器的界面马上发生了变化，如图 1.13 所示。

图 1.12　　　　　　　　　　　　图 1.13

在修改代码的过程中，我们并没有重新部署项目，只是简单地做了保存操作。这个特性在开发过程中是非常高效的，有点类似前端的 webpack 实现的热重载功能，对开发者非常友好，也提高了开发效率。

注意：本章 Demo 都是在 Android 和 iOS 模拟器上运行的。关于如何在真机上运行以及如何发布到 Android 应用市场和 App Store，笔者会在后续章节以及实战篇中进行介绍。

1.5　Flutter 升级

可以通过在项目的根目录中（包含 pubspec.yaml 的目录）输入命令"flutter upgrade"对 Flutter 进行升级，这里还需要介绍一下 Futter channels 相关的概念。

Flutter channels 有 4 种 release channel，即 stable、beta、dev、master，

官方推荐使用 stable channel，我们可以通过在命令行中输入"flutter channel"命令查看当前的 Flutter channels 情况，如图 1.14 所示。

图 1.14

图 1.14 中星号（*）的位置在 beta 上，表示当前 channel 为 beta。但根据官方推荐，我们可以考虑选择切换到 stable 上。那么怎样进行切换呢？可以输入以下命令：

```
flutter channel stable
```

切换完成之后，则 stable 为被选中状态，如图 1.15 所示。

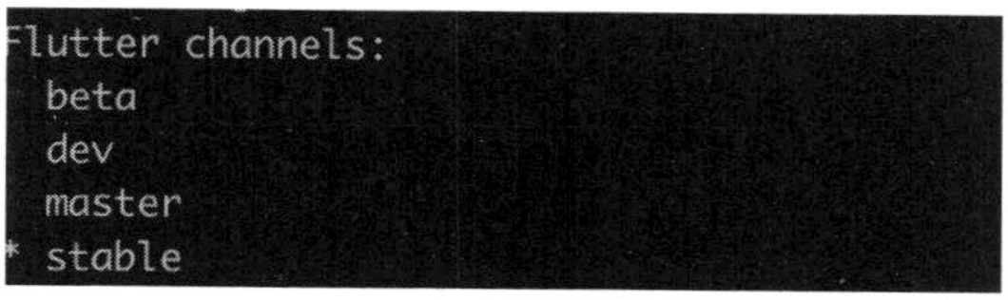

图 1.15

这里还需要注意 pubspec.yaml，这个文件掌管着 Flutter 所需要的依赖包，有点类似 Node.js 里的 package.json。这些 Flutter 依赖包有内置的，还有别的开发者发布的。我们可以通过"flutter packages get"命令获取依赖包列表，还能通过"flutter packages upgrade"命令升级这些依赖包。

本章小结

本章主要介绍了 Flutter 是什么，并对其基本的架构体系进行了学习。此外，本章还介绍了 Flutter 开发环境的搭建，模拟器的安装，第一个 Demo 的演示，Flutter 的升级方式等。接下来，第 2 章会介绍 Dart 语言，希望读者对 Dart 语法有一些初步的了解，然后从第 3 章开始，我们正式开始学习 Flutter 的具体开发内容。

第 2 章

Dart 语言入门

Dart 是谷歌发布的一种开源编程语言，初期目标是成为下一代 Web 开发语言，目前已经可用于全平台开发。Dart 又是面向对象的编程语言，本章所有的代码都可以在本书源码目录的 chapter2 文件夹里查到。

本章并不是专门介绍 Dart 语法的教程，并且假设读者已经有了一些其他语言的编程基础，所以，会跳过一些在计算机语言上非常基础的东西，比如常用的运算符和流程控制语句等。如果你已经对 Dart 语言有了深入的了解，可以直接跳过本章。

Dart 语言和 Java、JavaScript（JS）很像，从语法上来看是两者的结合体，目前最新的版本是 Dart 2.0 正式版。本书中所有的例子都会基于 Dart 2.0 开发。

2.1 应用场景

Flutter 适用于以下的场景：

（1）Web 开发

（2）跨平台移动应用开发（Flutter）

（3）脚本或服务端开发

本书介绍的 Dart 是用于开发 Flutter 的，因此不会介绍 Web 和服务端中的使用场景。

2.1.1　SDK 安装和升级

在 Dart 的官方网站上，根据自己的需要选择 SDK 平台，支持"command-line"命令行模式。笔者认为学习一门语言还是从命令行开始，手动敲代码比较好。

在 Homebrew 官方网站下载 Homebrew，然后执行以下代码：

```
brew tap dart-lang/dart
brew install dart -devel
```

安装完成后，可以执行"brew info dart"命令验证是否安装准确。

注意：本章中介绍的仅仅是 Dart 的 SDK 安装方式，和第 1 章介绍的 Flutter SDK 安装方式还是有区别的，本章不涉及 Flutter 创建的项目，因此只需要安装 Dart SDK 并且在 VSCode 中快速编写和运行代码即可。

2.1.2　编写一个 HelloWorld 并运行

大多数语言都是把 main 方法作为程序入口，Dart 也不例外。相比 Java 而言，Dart 简练了许多。代码如下所示：

```
void main() {
  print("Hello World!");    // 控制台打印
}
```

在代码写完之后，执行命令"dart hello world.dart"，稍等片刻就会出来结果。没错，在控制台输出的就是：Hello World!

2.2 变量与常量

掌握变量与常量是任何语言的基础，下面就来简单介绍一下 Dart 里面是怎样定义变量与常量的。

2.2.1 变量

使用 var 声明一个变量（这点和 JavaScript 声明变量的方式很像），然后可以给变量设置不同的值。在未初始化时，默认为 null。

需要重点注意的是，Dart 是一门强类型语言，在第一次赋值时，如果已经确定了是字符串类型，则不能更改为别的类型。而在 JavaScript 中，var a = 'hello dart'是没问题的，把 a 重新赋值，比如改成 a = 123，这样也是可以的，因为 JavaScript 是弱类型语言。如果真的想改变 Dart 中的字符串类型，可以使用 dynamic 关键字，后面的小节中会具体介绍。

2.2.2 常量

如果想设置一个常量，即让一个变量不可变，那么可以使用 final 或 const 进行修饰，例如：

```
var number;
number = 15;
print('小明的年龄是 $number 岁');

number = '20';        // 注意，变量类型是可以变的
print(number);

final c = 30;         // final 修饰的变量只能被声明一次
print(c);

const d = 50;
print(d)
```

那么 const 和 final 的区别是什么呢？

const 变量是一个编译时常量，final 变量在第一次使用时被初始化（const 变量是隐式的 final），实例变量可以是 final，但不能是 const。常量如果是类级别的，可以使用 static const，例如：static const PI = 3.1415。

2.2.3 内置类型

Dart 内置了一些类型，具体如下：

（1）Number（数值型）

（2）String（字符串）

（3）Boolean（布尔型）

（4）List（列表）

（5）Map（键值对）

（6）Runes（符号字符）

（7）Symbols（标识符）

2.2.4 数值型

数值型分整型 int 和浮点型 double，可以用 num、int、double 声明。

num 声明的变量加入的是 int 型，还可以被改成 double 型，但是，反过来 int 声明的变量不能再赋值成 double 型。

比如，正确的做法：

```
num a = 10;
a = 30.2;
```

错误的做法：

```
int b = 20;
b = 15.5;
```

看一下源码，我们就可以理解了，如下所示：

```
abstract class int extends num
abstract class double extends num
```

2.2.5 数值型操作

（1）运算符： +、-、*、/、~/、%。

（2）常用属性：isNaN、isEven、isOdd 等。

（3）常用方法：round()、floor()、ceil()、toInt()、toDouble()、abs()。

举一些例子，如下所示：

```
var a = 15;
var b = 10;

print(a + b);
print(a - b);
print(a * b);
print(a / b);
print(a ~/ b);
print(a % b);

print(a.isEven);
print(a.isOdd);
print(b.isEven);
print(b.isOdd);

var c = 1.23;

print(c.floor());
print(c.round());
```

```
print(c.floor());
print(c.ceil());
print(c.toInt());
```

在 Dart 2.0 里，如果我们声明的是 double 型，且值是 int 型时，int 型会自动转成 double 型，例如：

```
double a = 10        // 打印出来是 10.0
```

也可以通过 API 进行转化，例如：

```
print(a.toDouble());
```

2.2.6 字符串

之前在写 HelloWorld 时已经接触过字符串，那么在 Dart 里字符串有哪几种创建方式呢？

（1）使用单引号、双引号创建字符串。

（2）使用三个引号或双引号创建多行字符串。

（3）使用 r 创建原始 raw 字符串。

2.2.7 字符串操作

（1）运算符：+、*、==、[]。

（2）插值表达式：${expression}。

（3）常用属性：length、isEmpty、isNotEmpty。

（4）常用方法：contains()、subString()、startsWith()、endsWith()、indexOf()、lastIndexOf()、toLowerCase()、toUpperCase()、trim()、trimLeft()、trimRight()、split()、replaceAll()。

从名字来看，是不是很多 Java 里有的方法，Dart 里也有呢？对于字符串的常用方法，我们不用死记硬背。使用的时候如果想不起来可以去官网查看怎样使用，或者在 IDE 里面输入之后也会提示。举一个例子：

```
var str1 = 'Hello';
var str2 = "Hello2";
var str3 = '''你好
        欢迎光临''';
var str4 = r'双11真给力';
var str5 = 'a|b|c|d|e|f|g';
var str6 = '小学一年级';

print(str1.toUpperCase());
print(str2.length);
print(str3.startsWith('你好'));
print(str4);
print(str5.split('|'));
print(str6.replaceAll('小', '大'));
```

注意：字符串里单引号里面嵌套单引号，或者双引号里面嵌套双引号，必须在前面加反斜杠进行转意，这个和 JavaScript 里的字符串很像。笔者推荐双引号里嵌套单引号混合使用。

另外，我们再提一下字符串拼接。在 Dart 里，可以单行拼接，也可以多行拼接，例如：

```
String a1 = 'ha' 'ha';           // 单引号空格拼接
String a2 = "ha" "ha";           // 双引号空格拼接
String a3 = 'ha' + 'ha';         // 单引号加号拼接
String a4 = "ha" + "ha";         // 双引号加号拼接

// 使用三引号方式拼接多行字符串
String a5 = '''
  Android 开发工程师
  前端开发工程师
  ''';

String a6 = """Android 开发工程师
  前端开发工程师""";
```

我们再来介绍一下插值表达式，即${expression}，看一个例子：

```
double iphone = 11000.0;
print('最新的iphone价格为$iphone');
```

2.2.8 布尔型

布尔型通常用在 if 条件判断语句里面。

（1）使用 bool 表示布尔型。

（2）布尔值只有 true 和 false。

（3）布尔值是编译时常量。

（4）可以在 debug 模式下通过 assert 断言函数判断布尔值。如果不为真，会引发异常并终止程序往下运行。在开发时非常有用。我们来看一个例子：

```
var a = '';
assert(a.isEmpty);

var nullValue;
assert(nullValue == null);
```

2.2.9 List 与数组

在 Dart 中，List 表示集合，其实和数组是同一个概念。创建方式如下所示：

```
// 创建 List
var list = [1, 2, 3];

// 通过构造方式创建 List
var list2 = new List();

// 创建一个不可变的 List
var list3 = const[1, 2, 3];
```

在实际开发过程中，我们经常需要访问 List 里面的内容，可以通过 list[] 方式来进行，比如：list[0]。有的时候也需要改变 List 里面的内容，可以这样写：

```
list[0] = 1;
```

但是如果试图改变一个不可变的 List，就会报错。

List 中的常用方法有：length()、add()、insert()、remove()、clear()、indexOf()、lastIndexOf()、sort()、sublist()、asMap()、forEach()、shuffle()。

下面通过一个例子来具体学习 List，如下所示：

```
var list = ['one', 'two', 'three'];
print(list.length);            // 获取长度
list.add('four');              // 增加一个元素
print(list);
list.remove('two');            // 移除一个元素
print(list);
list.insert(1, 'two');         // 在指定位置插入一个元素
print(list);
print(list.indexOf('one'));    // 获取元素所在位置
print(list.sublist(2));        // 去除前 2 个元素，并返回一个新的 List
list.forEach(print);           // 传入一个方法
list.shuffle();                // 随机打乱 List 里元素的顺序
print(list);
```

2.2.10 Map

在 Dart 中，Map 以 key-value（键值对）形式存储，键和值都可以是任何类型的对象，每个键只出现一次。

（1）通过直接声明方式创建一个 Map，里面写 key 和 value，并且用逗号隔开，例如：

```
Map game = {"name": "Switch", "company": "任天堂"};
```

（2）创建一个不可变的 Map，只需在 Map 前面加入 const，例如：

```
Map game = const{"name": "Switch", "company": "任天堂"};
```

（3）构造方式先声明，然后再赋值，例如：

```
Map game = new Map();
game['name'] = 'Switch';            // 赋值
game['company'] = '任天堂';
```

上面的代码还可以通过"变量名[key] = value"的形式来修改 Map 里面的元素，例如：

```
game['name'] = 'GameBoy';
```

我们也可以通过调用 remove 或 clear 方法来移除元素，例如：

```
game.remove('name')                 // 把 key 为 'name' 的元素溢出掉
game.clear();                       // 清空整个 Map 集合
```

2.2.11 dynamic 和 Object

在 Dart 里面，一切皆对象，而且这些对象的父类都是 Object。

当没有明确类型时，编译的时候会根据值明确类型，例如：

```
var name1 = 'abc';
Object name2 = 'def';
dynamic name3 = 'hij';
```

以上写法都没有问题，但是 Dart 不建议我们这么做。在实际开发中，我们应尽量为变量确定一个类型，这样做可以提高安全性，加快运行速度。如果不指定类型，则在 debug 模式下类型会是动态的，所以笔者推荐这样写，如下所示：

```
String a = 'abc';
```

使用 dynamic 时则是告诉编译器，我们不用做类型检查，并且知道自己在做什么。如果我们调用一个不存在的方法时，会执行 noSuchMethod() 方法。在默认情况下（在 Object 里实现）它会抛出"NoSuchMethodError"。举一个例子来说明一下，如下所示：

```
dynamic obj = '小张';
obj['age'] = 20;
```

上面的代码编译时可以通过，但实际运行中会抛出"NoSuchMethodError"异常。

为了在运行时可以对类型进行检测，我们可以使用 as 和 is 关键字，例如：

```
dynamic obj = <String, int>{};
if (obj is Map<String, int>) {
    obj['age'] = 20;
}
var map = obj as Map<String, int>;
```

2.3 运算符

计算机语言中常用的几个运算符我们就不说了，来说一下 Dart 里常用的和特有的几个运算符。

1. 三目运算符

三目运算符在别的计算机语言中也有，在 Flutter 里用得比较多，通常与 state 状态管理结合，用来判断组件的状态，其基本用法如下所示：

```
expr1 ?? expr2//如果expr1非空，则返回其值，否则返回expr2的值
```

我们来看一个具体的例子，如下所示：

```
int a = 20;
var val = a > 10 ? a : 0;
```

2. ~/ 除法，返回一个整数结果（取商）

具体如下所示：

```
var val = 12~/7;
print(val);      // 结果是：1
```

3. 级联操作符

有点类似一些语言的链式调用，在 Java 和 JavaScript 里用得比较多。在后续有关动画的章节中，笔者会做更多介绍。例如：

```
String s = (new StringBuffer()
    ..write('a ')
    ..write('b ')
    ..write('c ')
  .toString();
  print(s);      // a b c
```

4. as、is 与 is!

as：判断属于某种类型。

is：如果对象具有指定的类型，则为 true。

is!：如果对象具有指定的类型，则为 false。

在前面讲 dynamic 的时候我们已经用相关例子说明了，这里就不再阐述。

2.4 异常捕获

Dart 的异常捕获比 Java 还要强大，可以抛出任意类型的对象。抛出异常的方式如下所示：

```
throw Exception('我是异常')
```

有的时候，我们也需要捕获异常，确保程序的健壮性（鲁棒性），例如：

```
try {
    // 捕获特定类型的异常
} on AuthorizationException catch (e) {
    // 捕获特定类型的异常，但不需要这个对象
} on Exception {
    // 捕获所有异常
} catch (e) {
```

```
    // ...
} finally {
    // ...
}
```

2.5 函数Function

Dart 是一门面向对象的语言,所以函数也是对象,并且函数的类型是 Function。这点和 JavaScript 非常像。在 JavaScript 里面 Function 是可以作为参数传递的,在 Flutter 里也不例外,即函数可以分配给变量或作为参数传递给其他函数。比如我们可以定义一个函数,如下所示:

```
bool getName(name) {
    return name;
}
```

2.5.1 main 函数

每个应用程序都必须有一个顶层 main 函数,这一点和 Java 一样,它是应用程序的入口点。该 main 函数返回 void 并具有 List 的可选参数。

我们看一下 Flutter 的入口函数是怎么写的,如下所示:

```
void main() => runApp(MyApp());
```

在 Flutter 中,main 就是入口函数,此处即 MyApp。

2.5.2 可选参数

可选的命名参数,即不传这些参数也可以。

在定义函数时,使用{param1, param2, …}指定命名参数。例如:

```
void userSettings({int age, String name}) {
    // ...
}
```

在上面的函数中，我们可以传递 age、name 这两个参数，或者其中一个参数，甚至不传递也可以。

2.5.3 必传参数

有的时候，我们在调用函数时必须传入一些参数，这个时候就可以用到 @required 来修饰。使用 @required 有利于静态代码分析器进行检查，例如：

```
void userSettings({@required int age, @required String name}) {
    // ...
}
```

2.5.4 可选的位置参数

用[]把目标标记为可选的位置参数，例如：

```
void userSettings({int age, String name, [String interests]}) {
    if (interests != null) {
        print('兴趣爱好$interests')
    }
}
```

2.5.5 默认参数

默认值是编译时常量，在函数的参数后面使用"="为参数赋值。这个有一点像 JavaScript 里的 ES6 特性，例如：

```
void userSettings({int age = 21, String name = '小明'}) {
    // ...
}
```

2.5.6 函数作为参数传递

函数可作为参数进行传递，即把一个函数作为一个参数传递给另外一个函数，例如：

```
void printItem(String item) {
  print(item);
}

var users = ['小明', '小王', '小张'];
uers.forEach(printItem);
```

2.5.7 函数作为变量

函数也可以直接赋值给一个变量，并且把这个变量作为函数来调用，例如：

```
var say = (name){
  print(name);
};
say('过年了');
```

2.6 异步编程

Dart 和 JavaScript 的共同点是——单线程，同步代码会阻塞程序。因此程序里能看到大量的异步操作，它是用 Future 对象来执行相关操作的，并且在 async 函数使用 await 关键字时才被执行，直到一个 Future 操作完成。Future 支持链式操作的方式，可以按顺序执行异步函数。

2.6.1　Future 是什么

一个 Future 是一个 Future 自身的泛型 Future<T>对象，它表示一个异步操作产生的 T 类型的结果。如果结果的值不可用，Future 的类型会是 Future<void>，如果返回一个 Future 的函数被调用了，将会发生以下两件事。

1. 这个函数加入待完成的队列并且返回一个未完成的 Future 对象。

2. 当这个操作结束了，Future 对象返回一个值或者错误。

举一个例子，如下所示：

```
Future<int> future = getFuture();
future.then((value) => handleValue(value))
      .catchError((error) => handleError(error))
.whenComplete() => handlerComplete();
```

如果有前端经验的同学，可以理解为 Future 类似前端里的 Promise。我们在 future.then 中接收异步处理的结果，并根据业务需求做相应的处理。而 future.catchError 则用于在异步函数中捕获并处理错误。在有些业务场景下，无论异步任务的处理结果是成功还是失败，我们都需要再进行一些处理，这时候就可以使用 Future 的 whenComplete 进行回调。

2.6.2　async 和 await

当遇到有需要延迟的运算（async）时，将其放到延迟运算的队列（await）中去，把不需要延迟运算的部分先执行完，最后再来处理延迟运算的部分。但是，要使用 await，就必须在有 async 标记的函数中运行，否则这个 await 会报错。

其实这和 JavaScript 也很像，async 和 await 是 Future 的语法糖（Syntactic sugar），解决了回调地狱（Callback Hell）的问题。

回调地狱是什么？这个要结合具体业务场景来说。比如"业务 1→业务 2→业务 3"，这 3 个任务都是异步的，那么业务 1 成功之后调用业务 1 的 then 回调，然后在业务 1 的 then 里面再调用业务 2，依此类推就形成了"回调地狱"。回调地狱多了之后，我们的代码会非常难看，因为需要大量的缩进。我们举一个实际的例子，如下所示：

```
step1('step1').then((step1Result){
   step2(step1Result).then((step2Result){
      step3(step2Result).then((step3Result){
         // step4
         //    step5
         //       step6
      });
   });
});
```

前面说了 async 和 await 是 Future 的语法糖，我们在使用过程中给人的感觉是在调用同步的代码。为了解决这种回调地狱，我们对以上代码结构进行一些调整，然后来看看修改后的代码，如下所示：

```
steps() async {
   try{
      String step1Result = await step1('step1');
      String step2Result = await step2(step1Result);
      String step3Result = await step3(step2Result);
      // step4
      // step5
      // step6
   } catch(e){
      print(e);
   }
}
```

从上面修改后的代码我们可以看出，await 必须被包裹在 async 里面，如果没有 async 就会报错。关于 async 和 await 我们就先介绍到这里，这部分内容需要读者重点掌握，后面 Flutter 网络请求耗时的操作会出现大量的 async/await。对于异步编程的概念我们先在本章了解一个大概，后续章节中我们会介绍异步编程里的重要概念，即 isolate 和 event loop。

2.6.3 继承、接口实现和混合

继承（extends）和接口实现（implements）这两个功能与 Java 比较相似，我们来回顾一下。

1. 继承（extends）。首先，明确一点，Flutter 中的继承也是单继承。Flutter 继承一个类之后，子类可以通过@override 来重写超类（父类）中的方法，也可以通过 super 来调用超类中的方法。需要说明的是，构造函数不能被继承。另外，由于 Flutter 没有 Java 中的公有和私有访问修饰符，因此，可以直接访问超类中的所有变量和方法。

2. 接口实现（implements）。Flutter 是没有接口（interface）关键字的，但是 Flutter 中的每个类都是一个隐式的接口，这个接口包含类里的所有成员变量和定义的方法。当类被当作接口使用时，类中的方法就是接口中的方法，它需要在子类里重新被实现。在子类实现的时候要加@override，我们看一段代码：

```
abstract class CanFixComputer {
  void fixComputer();
}

class CanProgramming {
  void programming(){

  }
}

class SoftwareEngineer extends Engineer implements CanFixComputer, CanProgramming {
  @override
  void fixComputer() {
    print('软件工程师修电脑');
  }
```

```
  @override
  void programming() {
    print('码农正在写代码');
  }
}
```

在上述代码中,如果 SoftwareEngineer 这个类不重写 programming 方法,则会报错,这一点和 Java 是一致的。

在介绍了继承和接口实现之后,我们来看一下 Dart 为我们新增的混合(mixins,也可以理解为混入)语法特性,它的作用是在类中混入其他功能。mixins 最早来自 Lisp 语言。说得更直白一点,在面向对象的语言中,mixins 是一个可以把自己的方法提供给其他类使用,但却不需要成为其他类的父类的类,它以非继承的方式来复用类中的代码。要使用 mixins,则要用关键字 with 来复用类中的代码,我们看一段代码:

```
abstract class CanFixComputer {
  factory CanFixComputer._() {
    return null;
  }

  void fixComputer() {
    print('软件工程师修电脑');
  }
}

abstract class CanProgramming {
  factory CanProgramming._() {
    return null;
  }

  void programming() {
    print('码农正在写代码');
  }
}
```

```dart
abstract class Worker {
  void doWork();
}

class Teacher extends Worker {
  void doWork() {
    print("老师在上课");
  }
}

class Engineer extends Worker {
  void doWork() {
    print("工程师在工作");
  }
}

class SoftwareEngineer extends Engineer with CanFixComputer, CanProgramming {
  @override
  void fixComputer() {
    print('软件工程师修电脑');
  }

  @override
  void programming() {
    print('码农正在写代码');
  }
}

class ITTeacher extends Teacher with CanFixComputer {
  @override
  void fixComputer() {
    print('IT教师修电脑');
  }
}

main() {
  ITTeacher()
```

```
    ..doWork()
    ..fixComputer();

  SoftwareEngineer()
    ..doWork()
    ..fixComputer()
    ..programming();
}
```

对于上述代码，我们同时运用了 extends 和 mixins，把 CanFixComputer 和 CanProgramming 混入到了 SoftwareEngineer 里面。

那么，如果同时使用 extends、implements 和 mixins，并且@override 的方法都一样，那么执行的优先级会怎样呢？我们看一个例子：

```
class Cat {
  void show() {
    print("小猫");
  }
}

class Bird {
  void show() {
    print("小鸟");
  }
}

class Owner {
  void show() {
    print("主人");
  }
}

class Person1 extends Owner with Cat, Bird {
//  void show() {
//    print("主人养了猫和小鸟");
//  }
```

```
}

class Person2 extends Owner with Cat implements Bird {
  // void show() {
  //   print("主人养了猫和小鸟");
  // }
}

main() {
  Person1()..show();
  Person2()..show();
}
```

如果放开上面代码中的注释，那么最先执行的是类本身的方法，如果注释之后，则执行优先级顺序是 mixins→extends→implements。读者可以多尝试。

2.6.4 泛型

Dart 中的泛型和 Java 中的很相似，比如 List<E>。用尖括号括起来的就是泛型的写法。在 List<E> 中，这个 E 代表泛型的类型，且不一定要用 E 表示，还可以用 T、S、K 等表示。举一个很简单的例子：

```
main() {
   List animals = new List<String>();
   animals.addAll(['小猫','小狗','小鸟']);
}
```

在上述代码中，我们用了 List 的集合来存放各种小动物，并且指定了 List<String> 的泛型，这样就表示了在这个定义的 List 中只能存放字符串类型。如果添加数字类型，比如 animals.add(123)，代码运行时就会报错。

那么，看了上面的例子之后，想一下，我们为什么要用泛型呢？从上面例子中的代码来看，指定了 List<String>的泛型，就可以使代码的逻辑控制更严谨，有效地对程序做类型检查。泛型的另一个好处是有效减少重复代码，并且在多种类型之间定义同一个接口实现。比如下面的写法是不用泛型的情况：

```
// 不用泛型，存储Object
abstract class ObjectData {
  Object getByKey(String key);
  void setByKey(String key, Object value);
}

// 不用泛型，存储String
abstract class StringData {
  String getByKey(String key);
  void setByKey(String key, String value);
}
```

上面的代码在使用泛型之后，则可以不再为每一种类型单独编写代码，如下所示：

```
abstract class Data<T> {
  T getByKey(String key);
  void setByKey(String key, T value);
}
```

Dart 泛型里面也可以通过 extends 限制参数的类型，比如下面的例子：

```
class Animal {}
class Cat extends Animal {}
class Bird extends Animal {}

class NewAnimal<T extends Animal> {
  String toString() => "创建一个新的小动物：'Foo<$T>'";
}
```

```
main() {
  var cat = NewAnimal<Cat>();
  var bird = NewAnimal<Bird>();
  var animal = NewAnimal();
  print(cat);
  print(bird);
  print(animal);
}
```

本章小结

本章学习了 Dart 的一些语法基础，并且在讲解过程中与 Java 和 JavaScript 进行了一些对比，便于读者快速掌握新语言。掌握 Dart 能为 Flutter 学习起到良好的铺垫作用，从下一章开始我们就正式学习 Flutter，并由此揭开 Flutter 的神秘面纱。

第 3 章

一切皆组件

你如果做过 Android 开发，那一定熟悉里面各种官方提供的布局，比如 LinearLayout、RelativeLayout 等。你也一定定制过一些自定义组件。如果你学过 Kotlin，并且用 XML 的 DSL 框架 Anko 声明 Android 组件，你会发现这与 Flutter 那种嵌套布局写法极为相似，如下所示：

```
// Kotlin 里 XML 的 DSL 举例
UI {
    // AnkoContext
    verticalLayout {
        padding = dip(30)
        button {
            // button
            id = R.id.todo_add
            textResource = R.string.add_todo
            textColor = Color.WHITE setBackgroundColor(Color.DKGRAY)
            onClick { _ -> createTodoFrom(title, content)}
        }
    }
}
```

从布局来看，官方把 Flutter 布局分为 Basic widgets，Single-child 和 Multi-child。在本章学习和实战开发的过程中，请记住一句关于 Flutter 的话：一切皆组件。我们来看一个简单的效果图，如图 3.1 所示。

图 3.1

图 3.1 可以看出这是嵌套关系组件，通过简单的思考和调整，我们把图 3.1 拆分成以下布局，如图 3.2 所示。

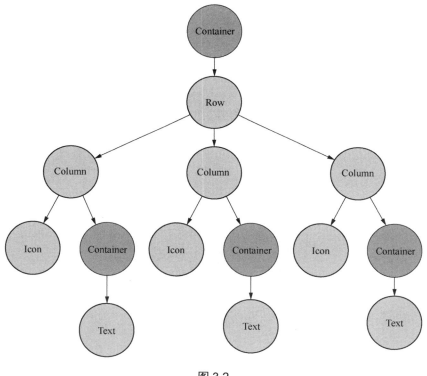

图 3.2

通过本章的学习，你可以了解一些常用的 Flutter 布局、对齐方式以及使用场景。当你看一眼图 3.3 就可以想出它是怎样的布局划分时，说明你已经掌握了 Flutter 的基本布局。

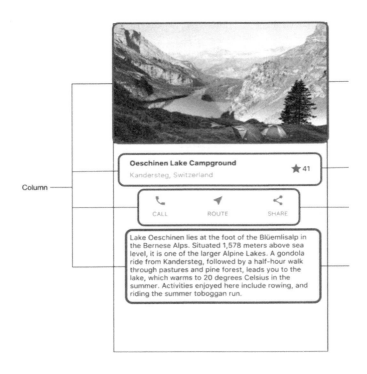

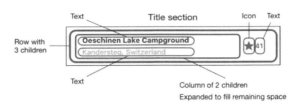

图 3.3

下面，让我们一起开始本章的学习吧。

3.1 基础组件（Basic widgets）

基础组件是组件的鼻祖，是必须要掌握的知识。在组件划分上，Container、Row、Column 在官方文档里属于基础组件。但是，Container 又属于单一子元素组件（Single-child）。而 Row，Column 又属于多子元素组件（Multi-child）。为了便于区分和记忆，本书约定基础组件是不能再拆分的原子组件，下面我们介绍一下几个非常常用的组件。

3.1.1 Text

Text 比较简单，基本的用法如下所示：

```
Text("hello");
```

当然，开发中仅仅用一些基本的文字输出，肯定还是不够的。在前端里的 CSS3 中，能通过样式修饰更多的文字显示效果，比如字体大小、阴影、颜色等。同样，在 Flutter 中 Text 的属性也很强大，我们来看一下常用的属性有哪些，如表 3.1 所示。

表 3.1

属性	取值
textAlign	对齐方式
maxLines	最大行数
textScaleFactor	缩放因子，默认值为 1.0
overflow	配合 maxLines 使用，超出最大行数时可以用省略号或渐变效果隐藏多余行数
style	TextStyle 对象，其属性有 color、fontFamily、background、fontSize、下画线等
textSpan	配合 Text.rich 使用，被包装成 RichText，可实现类似富文本的效果

其中常用的属性就是 style、textAlign。还有就是 maxLines 和 overflow 的结合使用，比如在一个详情页里，内容太长，这时候就可以结合这两种属性来使用。

关于 Text 的使用，后文很多例子都会涉及，这里就不一一举例了。

3.1.2 Icon

Icon 即图标，在 Android 中支持系统自带的图标，mipmap 文件夹里存放的就是 Icon 类型图片。

使用方式如下所示：

```
Icon(Icons.search: Colors.blue),
```

在 Flutter 系统自带的图标里，有如下几种：

```
Icons.access_alarm,
Icons.search,
Icons.movie
```

内置的 Icon 数量有很多，我们在 Android Studio 里可以根据输入提示，把它们展现出来，如图 3.4 所示。

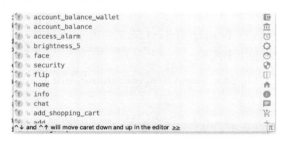

图 3.4

这些其实是矢量图标字体（iconfont），即这些图片的尺寸不管怎变，图标本身都是清晰的，且没有锯齿。我们可以像改变文字颜色一样改变图标颜色。但是，有的时候设计师不是这么想的。因为在大部分情况下，系统自带的图标肯定满足不了项目的需求。那么，在这个时候，可以考虑图标定制化。推荐大家使用阿里妈妈的矢量图标字体库，如图 3.5 所示。

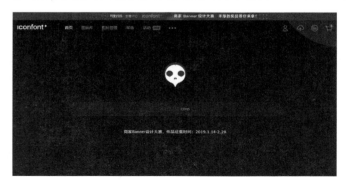

图 3.5

具体怎么导出自定义的图标字体，请读者按照官方网站的帮助文档来操作。Icon 除了使用图标字体以外，还支持传入图片，如下所示：

```
icon: "images/ic_main_tab_company_pre.png"
```

自定义图标字体的具体方式，会在本书最后的实战项目中有所描述。

3.1.3 Image

在 App 里，进行图片加载是不可或缺的。有了图片，可以让我们的 App 更加美观。我们来看一个基本的使用方式，如下所示：

```
// AssetImage 也能写成 Image.asset
Image(image: AssetImage("images/image_icon.png"), width: 50.0),
```

通过上述代码，我们实现了一个比较简单的本地图片加载流程。再来看一下，怎么从网络上加载一张图片，如下所示：

```
Image(
    image: NetworkImage("https://www.phei.com.cn/templates/images/img_logo.jpg"),
    width: 120.0,
)
```

除了上述方式，图片还支持通过 Image.file 和 Image.memory 的方式进行加载。可以看出，图片加载的方式有多种途径，可以是本地、网络、内存、文件等，而每种方式都继承了各自的 ImageProvider。

3.1.4 Button

Button 就是按钮。Flutter 中常用的按钮有以下几种：RaisedButton、FlatButton、IconButton、FloatingActionButton、OutlineButton 等。当然，我们也可以根据实际情况进行按钮的自定义。我们先来看一组按钮，如图 3.6 所示。

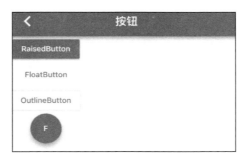

图 3.6

下面,我们依次来介绍这几个按钮。

RaisedButton:在点击时会带波纹效果,且有阴影,和 Android 里的 MD 风格按钮相似。代码如下所示:

```
RaisedButton(
    color: Colors.blue,
    child: Text("RaisedButton"),
    textColor: Colors.white,
    onPressed: () => {},
),
```

FlatButton:相对 RasiedButton 来说,它比较简洁,比较扁平,没有背景。代码如下所示:

```
FlatButton(
  textColor: Colors.blue,
  child: Text("FloatButton"),
  onPressed: () {
    print("FlatButton");
  },
),
```

OutlineButton:即带边框的按钮,当点击时,边框和背景颜色会呈现高亮状态。代码如下所示:

```
OutlineButton(
  textColor: Colors.blue,
  child: Text("OutlineButton"),
```

```
  onPressed: () {
    print("OutlineButton");
  },
),
```

FloatingActionButton：即 MD 风格按钮，一般来说一屏只有一个，用于分享、导航等。代码如下所示：

```
FloatingActionButton(
  child: Text("F"),
  onPressed: () {
    print("F");
  },
)
```

自定义按钮其实也比较简单，我们只需要把源码相关属性复写掉，就能得到想要的按钮效果。

完整的代码见 chapter3/flutter_widgets/lib/button_widget.dart。

3.1.5 FlutterLogo

顾名思义，FlutterLogo 即 Flutter 的 Logo，如图 3.7 所示。

图 3.7

我们可以通过传入 size 来改变 Logo 的大小，代码如下所示：

```
FlutterLogo(
  size: 100.0,
```

```
    colors: Colors.red,
)
```

小技巧：笔者在做具体项目时，比较喜欢在编写 FlutterLogo 布局的"占位图"时，先定下具体的图片尺寸（size），然后等整体布局完了之后，再把 FlutterLogo 替换成项目里具体的图片。

3.2 单一子元素组件（Single-child）

单一子元素组件（Single-child），包括 Container、Padding、Center、Align、FittedBox、AspectRatio 等。

3.2.1 Container

在 Flutter 中，使用最多的就是 Container，因此我们必须牢牢掌握跟它相关的知识。

先介绍一下 Container 的对齐方式（alignment），该属性接受 Alignment 对象。在其里面会传入两个参数，即 double x 和 double y，取值范围都在[-1, 1]之间，如图 3.8 所示。

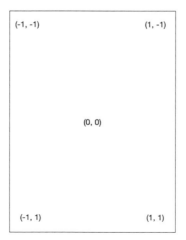

图 3.8

从图 3.8 我们可以看出，中间的位置 x 和 y 都为 0，它表示内容在 Container 里面正中间的位置。让一段文字内容在 Container 正中间的位置上，代码如下所示：

```
Container(
    color: Colors.green,
    alignment: Alignment(0.0, 0.0),
    child: new Text("Container"),
)
```

如果要让文字放在左下角，则只需把 Alignment(0.0, 0.0)改为 Alignment(-1, 1)即可。

有时候，专门去记位置的坐标比较麻烦，我们就需要用到 Flutter 提供的位置常量，如下所示：

```
Alignment.center == Alignment(0.0, 0.0)
Alignment.centerLeft == Alignment(-1.0, 0.0)
Alignment.topCenter == Alignment(0.0, -1.0)
Alignment.topLeft == Alignment(-1.0, -1.0)
Alignment.topRight == Alignment(1.0, -1.0)
Alignment.bottomCenter == Alignment(0.0, 1.0)
Alignment.bottomLeft == Alignment(-1.0, 1.0)
Alignment.bottomRight == Alignment(1.0, 1.0)
Alignment.centerRight == Alignment(1.0, 0.0)
```

3.2.2 Container 的约束

有的时候，我们需要约束容器所占据的大小和空间。在大部分情况下，可以通过 BoxConstraint 来构建完成，BoxConstraint 的属性如表 3.2 所示。

表3.2

属性	取值
minWidth	最小宽度
minHeight	最小高度
maxWidth	最大宽度
maxHeight	最大高度

我们先来看一段代码，如下所示：

```
child: Container(
  color: Colors.green,
  child: Text("Flutter 很棒"),
  constraints: BoxConstraints(
    maxHeight: 400.0,
    maxWidth: 300.0,
    minWidth: 200.0,
    minHeight: 200.0
  ),
)
```

有的时候，我们需要把储存在 child 中的 Container 扩展到最大，只需要加入一些约束条件即可，如下所示：

```
constraints: BoxConstraints.expend()
```

如果要设定容器的宽度和高度，也是有办法解决的，代码如下所示：

```
constraints: BoxConstraints.expand(width: 250.0, height: 100.0),
```

在 Container 里面，我们可以设置 margin（外边距）和 padding（内边距）的具体情况，比如：

```
padding: EdgeInsets.only()
margin: EdgeInsets.only()
```

其中 EdgeInsets 的写法有以下几种，如下所示：

```
EdgeInsets.symmetric（对称）
EdgeInsets.fromLTRB（位置）
EdgeInsets.All（所有）
```

我们用一张图来表示 margin 与 padding 的关系，如图 3.9 所示。

在 Container 里，仅仅限制一些组件的大小和位置是远远不够的，这时就有了 Decoration（装饰器）的概念。Decoration 的属性很强大，可以支持背景图线性或径向的渐变，也能支持边框、圆角、阴影等属性，如图 3.10 所示。

图 3.9

图 3.10

与图 3.10 所对应的代码如下所示：

```
Container(
    margin: EdgeInsets.only(top: 60.0, left: 80.0),
    constraints: BoxConstraints.tightFor(width: 300.0, height: 150.0),
    decoration: BoxDecoration(
      border: Border.all(width: 3, color: Color(0xffaaaaaa)),
      // 实现阴影效果
      boxShadow: [
        BoxShadow(
            color: Colors.black54,
            offset: Offset(2.0, 2.0),
            blurRadius: 4.0)
      ],
      // 实现渐变背景色，支持线性、径向渐变
      gradient: LinearGradient(
          colors: [Colors.red, Colors.blue, Colors.yellow]),
    ),
    transform: Matrix4.rotationZ(.3),
    alignment: Alignment.center,
    child: Text(
      "佩奇",
      style: TextStyle(color: Colors.white, fontSize: 30.0),
    ),
),
```

关于 Container 的使用，后续章节会有更多的例子，读者不需要死记硬背，只需要根据实际项目来使用，多试几次就会记住。

完整的代码见 chapter3/flutter_widgets/lib/container_widget.dart。

3.2.3 SingleChildScrollView

SingleChildScrollView 即滚动布局。从 SingleChildScrollView 的意思来看，这个组件是负责滚动的，里面只能嵌套一个组件，相当于 Android 里面的 ScrollView 布局。如果布局超出屏幕，是不能滚动的，就要在外层加上 SingleChildScrollView 组件才行，比如 Column 这种布局，代码如下所示：

```
//...
SingleChildScrollView(
  child: Column(...)
)
//...
```

SingleChildScrollView 可以设置滚动方向（水平或垂直），也能通过 reverse 属性设置阅读顺序（这个取决于语言）。

图 3.11 的布局，笔者就用到了 SingleChildScrollView。

图 3.11

为了让大家更好地掌握 Flutter 这门技术，本书后续的实战章节会做一个完整的案例供大家参考学习，此处就不详细展开了。

3.2.4 FittedBox

FittedBox 在官方的介绍中比较简短，但 FittedBox 在实际开发过程中还是比较有用的。其基本的定义是：负责对组件进行缩放和位置调整。

我们先来看一下 FittedBox 的主要属性，如表 3.3 所示。

表 3.3

属性	取值
fit	缩放方式
alignment	对齐方式

fit 指的是缩放本身占据 FittedBox 的大小，可以理解为 Android 里的缩放因子 scaleType，其默认值是 BoxFit.contain。也就是说，假如在 FittedBox 中给 fit 设置了 BoxFit.contain，那么当其子组件的宽度或高度被缩放到父容器限定的值时，就会被停止缩放。我们再来看一下其他属性的示例图，如图 3.12 所示。

图 3.12

大家可以根据项目的实际情况来设定 BoxFit 所对应的值。

完整的代码见 chapter3/flutter_widgets/lib/fittedbox_widget.dart。

3.2.5 FractionallySizedBox

FractionallySizedBox 看起来和 3.2.4 节中的 FittedBox 有点像，但实际运用起来还是有差异的。FractionallySizedBox 的用途是基于宽度缩放因子和高度缩放因子来调整布局大小的，大小有可能超出其父组件的设置。如果 FractionallySizedBox 中子组件设置了大小，它也不会起作用，而会被 FractionallySizedBox 的宽度缩放因子和高度缩放因子所覆盖。我们看一段具体的代码，如下所示：

```
Container(
  color: Colors.blue,
  height: 130.0,
  width: 130.0,
  child: new FractionallySizedBox(
    alignment: Alignment.topLeft,
    widthFactor: 1.5,
    heightFactor: 0.5,
    child: new Container(
      width: 50.0,     // 无效
      color: Colors.yellow,
    ),
  ),
),
```

在上述代码中，即使我们对 FractionallySizedBox 里的 Container 设置了宽度，也是不起作用的。Flutter 只会识别 FractionallySizedBox 中 widthFactor 和 heightFactor 所设置的值。

通过一张图，我们来详细了解一下 widthFactor 和 heightFactor，如图 3.13 所示。

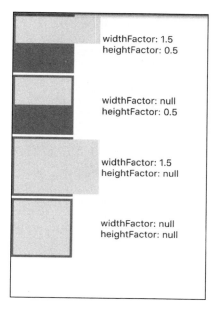

图 3.13

在图 3.13 中，我们为 widthFactor 和 heightFactor 分别设置了不同的值来展示其效果。完整的代码见 chapter3/flutter_widgets/lib/fractionallysized box_ widget.dart。

3.2.6 ConstrainedBox

ConstrainedBox 是一种有约束限制的布局，在其约定的范围内，比如最大高度、最小宽度，其子组件是不能逾越的。我们来看一个例子，代码如下所示：

```
ConstrainedBox(
  constraints: BoxConstraints(
    minWidth: 100.0,
    minHeight: 100.0,
    maxWidth: 250.0,
    maxHeight: 250.0,
  ),
  child: new Container(
```

```
            width: 300.0,
            height: 300.0,
            color: Colors.blue,
        ),
    ),
```

从例子中我们可以看出，在 ConstrainedBox 之中，主要是 constraints 在起作用，而且这个值不能为 null（空），是必须传入的。constraints 定义了最小宽度、最小高度、最大宽度、最大高度，我们看一下其属性列表，如表 3.4 所示。

表 3.4

属性	取值
minWidth	最小宽度
minHeight	最小高度
maxWidth	最大宽度
maxHeight	最大高度

完整的代码见 chapter3/flutter_widgets/lib/constrainedbox_widget.dart。

3.2.7 Baseline

通俗地讲，Baseline 是一种基线的对齐方式，它可以把不相关的几个组件设置在同一条水平线上进行对齐，如图 3.14 所示。

图 3.14

这一部分比较简单,笔者写了一些代码,然后做了相关的效果图,如图 3.15 所示。

图 3.15

可以看出,3 个控件都在底部对齐了,这就是 Baseline 的作用。需要注意的是,在代码中 baseline 和 baselineType 都不能为空,对应的代码如下所示:

```
body: new Row(
  mainAxisAlignment: MainAxisAlignment.spaceBetween,
  children: <Widget>[
    new Baseline(
      baseline: 100.0,
      baselineType: TextBaseline.alphabetic,
      child: new Text(
        '今天天气真好',
        style: new TextStyle(
          fontSize: 18.0,
          textBaseline: TextBaseline.alphabetic,
        ),
      ),
    ),
    new Baseline(
      baseline: 100.0,
      baselineType: TextBaseline.alphabetic,
      child: new Text(
        '适合晨练',
        style: new TextStyle(
          fontSize: 30.0,
          textBaseline: TextBaseline.alphabetic,
        ),
```

```
      ),
    ),
    new Baseline(
      baseline: 100.0,
      baselineType: TextBaseline.alphabetic,
      child: FlutterLogo(
        size: 100,
      ),
    ),
  ],
),
```

完整的代码见 chapter3/flutter_widgets/lib/baseline_widget.dart。

3.2.8　IntrinsicWidth 和 IntrinsicHeight

IntrinsicWidth 和 IntrinsicHeight 是官方不推荐使用的组件。它们存在一些性能上的问题，这里就不再描述。在一般情况下，这两种布局都有可以取代的方式。另外，笔者暂未发现必须使用这两种组件的使用场景。

3.3　多子元素组件（Multi-child）

多子元素组件（Multi-child），包括 Scaffold、AppBar、Row、Column、Stack、IndexedStack、ListView、GridView、Flow、Table、Flex、Wrap、CustomScrollView、CustomMultiChildLayout 等。接下来介绍其中比较常见的组件。

3.3.1　Scaffold

Scaffold 是基于 Material 库的一个与路由相关的、良好的"结构体"，它可以被认为是 Flutter 为我们提供的一个标准化的布局容器。我们经常可以在代码中看到它的身影，结构形式如图 3.16 所示。

图 3.16

代码结构如下所示：

```
Widget build(BuildContext context) {
  return Scaffold(
    AppBar: ...
    body: ...
    bottomNavigationBar: ...
    floatingActionButton: ...
    drawer: ...
    ...
  )
}
```

根据上述代码，我们可以了解到，Scaffold 这种"结构体"为我们很好地集成了 AppBar、floatingActionButton 等。这一节，我们先简单认识 Scaffold。它里面所包裹的那些组件，在后面的小节，笔者会逐步告诉大家怎么使用。

3.3.2 AppBar

AppBar，即 App 的顶部导航栏，用于控制 App 的路由、显示标题栏，以及显示右侧的一些操作栏。其绘制区域一般位于屏幕的顶端，如图 3.17 和图 3.18 所示。

图 3.17

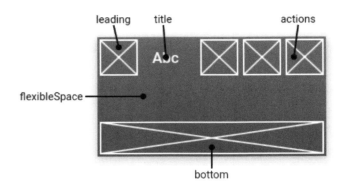

图 3.18

从图 3.18 可以看出，leading 区域默认是隐藏的，但是如果有左侧滑栏时则会显示。并且，当这个界面是上一个界面跳转过来的时候，就存在"上一级路由"，在界面上就会有返回按钮。我们来看一个例子，代码如下所示：

```
AppBar(
  title: Text('本地优惠'),
  actions: <Widget>[
    IconButton(
```

```
        icon: Icon(Icons.playlist_play),
        tooltip: 'tooltip1',
        onPressed: (){},
      ),
      IconButton(
        icon: Icon(Icons.playlist_add),
        tooltip: 'tooltip2',
        onPressed: (){},
      ),
      IconButton(
        icon: Icon(Icons.playlist_add_check),
        tooltip: 'tooltip3',
        onPressed: (){},
      ),
    ],
)
```

在一般情况下，我们用上面的代码基本可以实现大部分需求。但是，在有的情况下，我们还是需要自定义。比如，左侧不一定是返回按钮，有可能是别的图标和点击功能。这个时候，我们可以重写 leading 属性，代码如下所示：

```
leading: IconButton(
    icon: new Icon(Icons.face),
*   onPressed: () {
}),
```

这样，我们就可以自定义左侧的图标和点击功能了。其实 title 也不一定只能传 Text 组件，还能根据实际需要传入别的组件。

完整的代码见 chapter3/flutter_widgets/lib/appbar_widget.dart。

3.3.3 Row 和 Column

Row 和 Column 在 Flutter 里的布局是需要我们重点掌握的内容，它们都属于线性布局。从 Android 层面来理解，类似的布局是 LinearLayout；从前端的角度来理解，有点像标签的使用。

需要注意的是，Column 是不支持滚动的，如果需要实现滚动功能，则需要考虑使用 ListView。

我们先来看一下 Row。Row 是一个多子元素组件，用于在水平方向上放置并显示子组件，Row 的基本属性如表 3.5 所示。

表 3.5

属性	取值
children	传入子组件的数组
crossAxisAlignment	子组件在纵轴方向上的对齐方式
mainAxisAlignment	子组件在水平方向上的对齐方式
textDirection	布局顺序，一般情况下从左到右
mainAxisSize	max，表示尽可能多地占用水平方向上的位置，min 则反之

根据 Row 的属性，我们举一个例子，代码如下所示：

```
Row(
  textDirection: TextDirection.rtl,    // 从右到左
  children: <Widget>[
    Container(
      width: 100,
      height: 100,
      color: Colors.blue,
      alignment: Alignment.center,
      child: Text("A",
        style: new TextStyle(color: Colors.white, fontSize: 25.0)),
    ),
    Spacer(
      flex: 1,
    ),
    Container(
      width: 100,
      height: 100,
      color: Colors.blue,
      alignment: Alignment.center,
      child: Text("B",
```

```
          style: new TextStyle(color: Colors.white, fontSize: 25.0)),
    ),
    Spacer(
      flex: 1,
    ),
    Container(
      width: 100,
      height: 100,
      color: Colors.blue,
      alignment: Alignment.center,
      child: Text("C",
          style: new TextStyle(color: Colors.white, fontSize: 25.0)),
    ),
  ],
)
```

效果如图 3.19 所示。

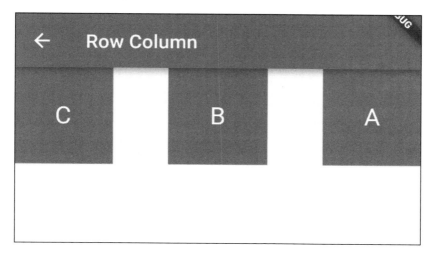

图 3.19

我们再来看一下 Column 组件。它其实和 Row 组件很相似，区别仅仅在于它是在垂直方向上放置多个组件。这里就不过多阐述。但是需要注意的是，在 Row 和 Column 里 Cross Axis 和 Main Axis 是不一样的，如图 3.20 所示。

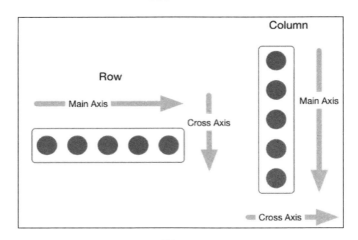

图 3.20

完整的代码见 chapter3/flutter_widgets/lib/row_column_widget.dart。

3.3.4 ListView

Flutter 中的 ListView 和 Android 中的 ListView、RecycleView 有些相似，作用都是可滚动项的线性列表，里面存放着相关组件的集合。在一般情况下，这些组件的结构都是具有重复性的，即每个 item 的结构基本相同。

ListView 的构建方式有以下几种：

（1）ListView

（2）ListView.builder

（3）ListView.custom

（4）ListView.separated

我们先看一下 ListView 的构建方式，代码如下所示：

```
ListView(
    padding: const EdgeInsets.all(10.0),
    itemExtent: 30.0,
    children: <Widget>[
      Text('A'),
```

```
        Text('B'),
        Text('C'),
        Text('D'),
        Text('E')
    ],
),
```

最终的效果图如图 3.21 所示。

```
A
B
C
D
E
```

图 3.21

这是最传统的创建方式。需要注意的是，里面有一个参数 itemExtent 被设置成了 30.0，在这里表示该 ListView 的子项的高度是 30.0。如果 ListView 本身是水平滚动的（scrollDirection: Axis.horizontal），则子项的宽度为 30.0。在设置了 itemExtent 的情况下，代码运行效率会比不设置的情况高，这是因为不需要做动态计算了。ListView 还有几个属性也是需要我们掌握的，如表 3.6 所示。

表 3.6

属性	取值
shrinkWrap	是否根据子组件的高度来设置 ListView 的高度，默认为 false
RepaintBoundary	当 addRepaintBoundaries 为 true 时，避免列表项滚动时重绘
AutomaticKeepAlive	当 addAutomaticKeepAlives 为 true 时，则 ListView 被滑出的区域不会被回收

ListView.builder 被用于创建重复的子项布局，将上面的部分代码改写一下就可以做到，改写后的代码如下所示：

```
ListView.builder(
    padding: const EdgeInsets.all(10.0),
    itemExtent: 30.0,
    itemCount: 5,      // 可以不传
    itemBuilder: (context, position) {
```

```
      return ListItem();
    },
  ),
```

ListView 是懒加载的，所以可以不指定 itemCount。列表有可能是无限长的，而 itemBuilder 传入的类型是 IndexedWidgetBuilder，只会返回一个组件。当我们滚动到具体位置因子时，就会创建列表项。

我们再来看一下 ListView.separated。它的作用是"分割"，即在列表子项中夹杂其他项。代码的基本结构如下所示：

```
//...
ListView.separated(
  itemCount: itemCount,
  itemBuilder: (context, position) {
    return ListItem();
  },
  separatorBuilder: (context, position) {
    return SeparatorItem();
  },
),
//...
```

separatorBuilder 和 itemBuilder 基本相似。我们在实际项目实践当中，有可能需要用到分割线来分割列表的每一项 item，这个时候 ListView.separated 就派上用处了。然而，在 separatorBuilder 里面，不一定只能传入分割线，也可以传入其他的东西，比如图片。我们看一个效果，如图 3.22 所示。

图 3.22 所对应的代码如下所示：

```
body: ListView.separated(
  itemCount: 20,
  itemBuilder: (BuildContext context, int index) {
    return ListTile(title: Text("列表项$index"));
  },
  separatorBuilder: (BuildContext context, int index) {
    return Align(
      alignment: Alignment.centerLeft,
      child: FlutterLogo(),
```

```
          );
        },
      ),
```

图 3.22

最后,来说一下创建方式 ListView.custom。它可以通过 SliverChildListDelegate 来接收 IndexedWidgetBuilder,并且为 ListView 生成列表项,从而实现自定义功能。

以上便是 ListView 的各种创建方式。我们再来探索一下 ScrollPhysics。这个属性是在 ListView 中的 physics 上面设置的,包括 NeverScrollablePhysics (不滚动效果)、BouncingScrollPhysics(iOS 效果)、ClampingScrollPhysics (Android 效果)、FixedExtentScrollPhysics(固定范围的滚动效果)等。我们来看一个 FixedExtentScrollPhysics 相关的例子,效果如图 3.23 所示。

图 3.23

其实它的代码也非常简单,如下所示:

```
body: ListWheelScrollView(
    controller: fixedExtentScrollController,
    physics: FixedExtentScrollPhysics(),
    itemExtent: 150.0,
    children: imgList.map((img) {
      return Card(
        child: Row(
          children: <Widget>[
            Image.network(
              img,
              width: 150.0,
            ),
            Text(
```

```
            '文字介绍',
          style: TextStyle(fontSize: 20.0),
        )
      ],
    ));
  }).toList(),
),
```

关于 ListView 的使用就介绍到这里,本节完整的代码见 chapter3/flutter_widgets/lib/listview_widget.dart(注:本例只适用于 Flutter 1.2 版本,1.7 版本暂不支持,此问题目前在官方的 issue 里还是 open 状态)。

3.3.5 GridView

和 ListView 相似,GridView 只不过是网格形式的表现,它有点像 Android 的 LayoutManager。如果 GridView 组件做得好、运用合理,那么 App 界面就会很漂亮,给人赏心悦目的感觉。需要重点注意的是,GridView 中的 gridDelegate 属性,其类型是 SliverGridDelegate,是一个抽象类,通过该类可以控制 GridView 的排列显示方式。

我们看一下 Flutter 官方的实现方式,代码如下所示:

```
GridView.count(
  primary: false,
  padding: const EdgeInsets.all(20.0),
  crossAxisSpacing: 10.0,
  crossAxisCount: 2,
  children: <Widget>[
    const Text('He\'d have you all unravel at the'),
    const Text('Heed not the rabble'),
    const Text('Sound of screams but the'),
    const Text('Who scream'),
    const Text('Revolution is coming...'),
    const Text('Revolution, they...'),
  ],
)
```

其中的 gridDelegate 在哪里呢？其实，在 GridView.count 中的构造函数里已经被传入了默认的 gridDelegate，我们进入源码就可以看到，如图 3.24 所示。

```
gridDelegate = SliverGridDelegateWithFixedCrossAxisCount(
  crossAxisCount: crossAxisCount,
  mainAxisSpacing: mainAxisSpacing,
  crossAxisSpacing: crossAxisSpacing,
  childAspectRatio: childAspectRatio,
),
```

图 3.24

源码里已经使用了 SliverGridDelegateWithFixedCrossAxisCount。细心的读者就会发现 SliverGridDelegateWithFixedCrossAxisCount 里有 4 个属性，我们来看一下它们的定义，如表 3.7 所示。

表 3.7

属性	取值
crossAxisCount	横向轴子元素的数量
mainAxisSpacing	横向轴之间的间距
crossAxisSpacing	子元素之间的间距
childAspectRatio	子元素的宽高比，比如 2.0 就表示宽度是高度的 2 倍

gridDelegate 还支持 SliverGridDelegateWithMaxCrossAxisExtent。GridView 也支持通过 GridView.builder 来创建，这里就不再赘述。

完整的代码见 chapter3/flutter_widgets/lib/gridview_widget.dart。

3.3.6 CustomScrollView

在实际应用里，布局情况是比较复杂的，一般一个界面不会只有一个滚动列表组件（ListView），很可能还有别的组件。比如 3.3.5 节提到的 GridView 组件，在大部分情况下，它是两个组件结合起来形成的一个滚动区域。这个时候，CustomScrollView 就有用处了。先来看一个效果图，如图 3.25 所示。

第 3 章 一切皆组件

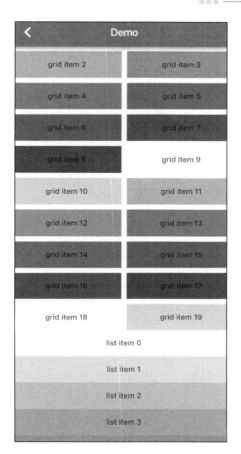

图 3.25

再来看一下图 3.25 所对应的代码，如下所示：

```
CustomScrollView(
  slivers: <Widget>[
    const SliverAppBar(
      pinned: true,
      expandedHeight: 250.0,
      flexibleSpace: FlexibleSpaceBar(
        title: Text('Demo'),
      ),
    ),
    SliverGrid(
      gridDelegate: SliverGridDelegateWithMaxCrossAxisExtent(
```

```
      maxCrossAxisExtent: 200.0,
      mainAxisSpacing: 10.0,
      crossAxisSpacing: 10.0,
      childAspectRatio: 4.0,
    ),
    delegate: SliverChildBuilderDelegate(
      (BuildContext context, int index) {
        return Container(
          alignment: Alignment.center,
          color: Colors.teal[100 * (index % 9)],
          child: Text('grid item $index'),
        );
      },
      childCount: 20,
    ),
  ),
  SliverFixedExtentList(
    itemExtent: 50.0,
    delegate: SliverChildBuilderDelegate(
      (BuildContext context, int index) {
        return Container(
          alignment: Alignment.center,
          color: Colors.lightBlue[100 * (index % 9)],
          child: Text('list item $index'),
        );
      },
    ),
  ),
],
)
```

在上述代码运行之后，只有一个大的滚动区域。有一些布局已经做了组件上的转换，代码如下所示：

```
AppBar -> SliverAppBar
GridView -> SliverGrid
ListView -> SliverFixedExtentList
```

注意：转换后的组件都是以"Sliver"开头的，其本身是不具备滚动特征的，但是当放在 CustomScrollView 中之后，则可以实现滚动的功能。

完整的代码见 chapter3/flutter_widgets/lib/customscrollview_widget.dart。

3.3.7 Flex

Flex 即弹性布局，该布局借鉴了前端里的 Flex 布局方式。用法也十分简单，我们可以在 Flex 中传入一些参数，其具体属性如表 3.8 所示。

表 3.8

属性	取值
direction	Axis.vertical 表示垂直方向，Axis.horizontal 表示水平方向
flex	弹性系数，大于 0 会按比例来分割，等于 0 不会扩展占用的空间

我们看一下图 3.26 的效果。

图 3.26

图 3.26 所对应的代码如下所示：

```
class _FlexWidgetState extends State<FlexWidget> {
  @override
  Widget build(BuildContext context) {
    return Scaffold(
      appBar: AppBar(
        title: Text("Flex"),
      ),
      body: Column(
        children: <Widget>[
          Container(
            height: 400.0,
            child: Flex(
              direction: Axis.vertical,
              children: <Widget>[
                Expanded(
                  flex: 1,
                  child: Container(
                    color: Colors.red,
                  ),
                ),
                Expanded(
                  flex: 2,
                  child: Container(
                    color: Colors.yellow,
                  ),
                ),
              ],
            ),
          ),
          Container(
            height: 120.0,
            child: new Row(
              mainAxisAlignment: MainAxisAlignment.spaceBetween,
              // 子组件的排列方式为主轴两端对齐
              children: <Widget>[
                Expanded(
                  flex: 2,
```

```
              child: Container(
                color: Colors.blue,
              )),
              Expanded(
                flex: 1,
                child: Container(
                  color: Colors.red,
                ))
            ],
          ))
        ],
      ));
    }
  }
```

在上面这个例子中，我们实现了 Flex 布局。这种布局还有一种方式，它通过在 Row 组件里设置两边对齐（mainAxisAlignment: MainAxisAlignment.spaceBetween）来实现。

完整的代码见 chapter3/flutter_widgets/lib/flex_widget.dart。

3.3.8 Wrap

Wrap 从字面意思来说是"包裹"（动词）的意思。在前端语义化和业务场景下，我们通常把具有相同点的布局整合在一个大的布局组件之内。之前笔者介绍过，可以使用 Row 和 Column 去包裹一些组件，因为 Row 和 Column 都是多子元素组件。Column 布局的情况可能还好，但是 Row 有的时候就会出问题，如图 3.27 所示。

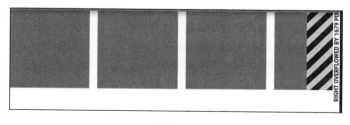

图 3.27

在图 3.27 中可以明显看到右侧有问题。在控制台中，也会有一些提示，如图 3.28 所示。

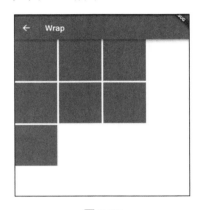

图 3.28

这种情况，是由 Row 中的控件在水平位置上"撑破"了屏幕所导致的。实际上，这种情况我们可以进行换行处理，即使用 Wrap 代替 Row，我们看一下修改后的效果，如图 3.29 所示。

图 3.29

图 3.29 所对应的代码如下所示：

```
Wrap(
  spacing: 4.0,
  runSpacing: 4.0,
  children: <Widget>[
    Container(
      width: 100,
      height: 100,
      color: Colors.blue,
    ),
    Container(
      width: 100,
      height: 100,
```

```
      color: Colors.blue,
    ),
    // 此处省略更多Container布局代码
  ],
)
```

我们来看一下 Wrap 里的常用属性，如表 3.9 所示。

表 3.9

参数	取值
spacing	水平方向间距
runSpacing	垂直方向间距

完整的代码见 chapter3/flutter_widgets/lib/wrap_widget.dart。

3.4 状态管理

在学习完 Flutter 的常用组件之后，我们对这些组件的使用场景有了一定的了解。但是，只有了解还是不够的，因为我们的数据是动态的，具备交互性，界面上的展示会根据用户的操作产生变化。这个时候，就需要对 Flutter 的状态管理机制有一定的了解。下面我们来看一下 Flutter 是如何管理组件状态的，如图 3.30 所示。

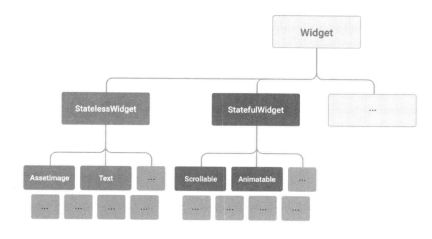

图 3.30

本章开头笔者就说过，Flutter 一切皆组件，而组件（Widget）主要被划分成 StatelessWidget 和 StatefulWidget 两大类，下面我们分别来介绍一下这两类组件。

3.4.1 Widget 树

我们在界面布局之后，会出现布局层级。不管这个界面是简单的，还是复杂的，都会有布局的嵌套层级，这样的布局我们称为 Widget 树，并且在 Widget 之间存在着"父子"关系。其实 Widget 只是起到"描述"的作用，真正在屏幕上显示给用户看的 UI 是 Element，Widget 只是描述了 Element 的数据配置，Widget 树的展现形式如图 3.31 所示。

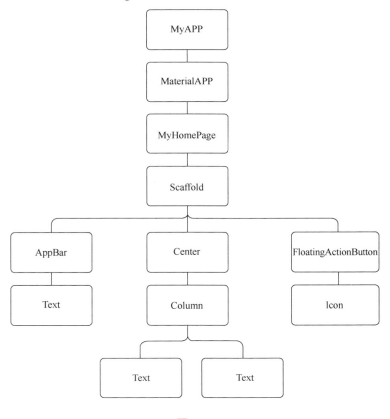

图 3.31

如果一个 Widget 存在子 Widget，那么上一层的 Widget 就是父 Widget，父 Widget 里面的则是子 Widget。在实际开发中，我们会遇到复杂的布局情况，这时候，不妨先在脑海里想一下，这一个布局块的 Widget 树是怎么样的。这样，接下来的操作实施就更加清晰、容易了。

3.4.2 Context

Context 则表示组件上下文的意思。Context 对应的是，构建 Widget 树结构中具体某一个 Widget 的位置引用，并且它被视为 Widget 的一部分，而每一个 Context 只对应一个 Widget。如果一个 Context 对应的是父 Widget A，则 Widget A 对应的 Context 也包含了子 Widget 的 Context。通过 Context 可以遍历和查找当前 Widget 树，Context 之间也是关联在一起的，它们组成一棵 Context 树。Context 示意图如图 3.32 所示。

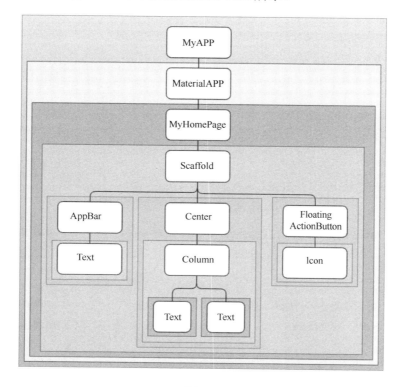

图 3.32

3.4.3 StatelessWidget

StatelessWidget 即无状态的 Widget，它无法通过 setState 设置组件的状态。其写法通过继承 StatelessWidget，然后重写 build 函数来实现。对于其内部属性，应该被声明为 final，以防止被意外改变，例如：

```
class MyStatelessWidget extends StatelessWidget {
   MyStatelessWidget({
      Key key,
      this.parameter,
   }): super(key:key);

   final parameter;

   @override
   Widget build(BuildContext context){
      return new ...
   }
}
```

上例中的 final parameter 参数只能传递一次，之后就无法被修改了。它的生命周期非常简单，就两步：（1）初始化；（2）通过 build 进行渲染。

3.4.4 StatefulWidget

StatefulWidget 即有状态的 Widget。当我们创建一个 StatefulWidget 组件时，它同时也创建了一个 State 对象，并且 StatefulWidget 通过与 State 关联可以达到刷新 UI 的目的。我们看一下官方的一个图示，如图 3.33 所示。

和 React/Vue 中的 Reactive 思想相似（但还是有区别的），StatefulWidget 只需要调用 setState(…) 方法，而不需要调用类似 Android 里的 textView.setText(…)方法。某些组件在其生命周期里，内部数据会发生变化，基于这种情况，我们应该考虑使用 StatefulWidget。

图 3.33

3.4.5　StatefulWidget 的组成

StatefulWidget 由两部分组成,第一部分为主体部分,代码如下所示:

```
class MyStatefulWidget extends StatefulWidget {
    MyStatefulWidget({
        Key key,
        this.color,
    }): super(key: key);

    final Color color;

    @override
    _MyStatefulWidgetState createState() => new _MyStatefulWidgetState();
}
```

在主体部分里 Widget 继承了 StatefulWidget 的内容。在主体部分中创建的变量是无法变更的,第二部分代码如下所示:

```
class _MyStatefulWidgetState extends State<MyStatefulWidget> {
    ...
    @override
    Widget build(BuildContext context){
        ...
    }
}
```

MyStatefulWidgetState 是以""开头的,表示该类是"私有的",可以通过 widget.{name of the variable}格式访问第一部分里定义的变量,比如 widget.color。

3.4.6 State

State 是对 StatefulWidget 的行为和布局的描述，和 StatefulWidget 存在一一对应的关系。前文说到，有的时候 StatefulWidget 的组件需要改变其界面的表现形式，这个时候就可以通过 setState(…)来改变。Flutter 为我们做的服务功能非常全面，改变的过程只需要由 Framework 层控制，从而达到更新 UI 的目的。

3.4.7 State 生命周期

在学习 React 和 Android 时，我们都会接触生命周期的概念。Flutter 也一样，Framework 层为 StatefulWidget 对应的 State 定制了生命周期的回调方法，需要读者认真阅读并掌握。State 完整的生命周期如图 3.34 所示。

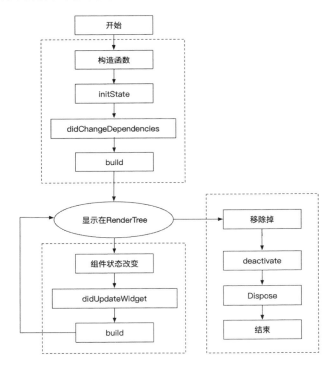

图 3.34

那么，这些回调方法分别是在什么时候调用的呢？这部分内容千万不要死记硬背，笔者会通过例子并结合上面的生命周期原理图来说明。

我们通过"flutter create flutter_lifecycle"创建一个生命周期的项目，并且里面默认的代码就是官方提供的计数器的代码，然后我们在 main.dart 里补上生命周期的方法并打印出来，完整代码如下所示：

```dart
import 'package:flutter/material.dart';
import 'package:flutter_lifecycle/next_page.dart';

void main() => runApp(MyApp());

class MyApp extends StatelessWidget {
  @override
  Widget build(BuildContext context) {
    return MaterialApp(
      title: 'Flutter Demo',
      theme: ThemeData(
        primarySwatch: Colors.blue,
      ),
      home: MyHomePage(title: 'Flutter Demo Home Page'),
    );
  }
}

class MyHomePage extends StatefulWidget {
  MyHomePage({Key key, this.title}) : super(key: key);

  final String title;

  @override
  _MyHomePageState createState() => _MyHomePageState();
}

class _MyHomePageState extends State<MyHomePage> {
  int _counter = 0;

  @override
```

```
void initState() {
  super.initState();
  print('initState');
}

void _incrementCounter() {
  setState(() {
    _counter++;
  });
}

@override
Widget build(BuildContext context) {
  print('build');
  return Scaffold(
    appBar: AppBar(
      title: Text(widget.title),
      actions: <Widget>[
        new IconButton(
          icon: new Icon(
            Icons.navigate_next,
            color: Colors.white,
          ),
          onPressed: () {
            Navigator.push(
              context,
              MaterialPageRoute(
                // fullscreenDialog: true,
                builder: (context) => NextPageWidget(),
              ),
            );
          },
        )
      ],
    ),
    body: Center(
      child: Column(
        mainAxisAlignment: MainAxisAlignment.center,
```

```
          children: <Widget>[
            Text(
              'You have pushed the button this many times:',
            ),
            Text(
              '$_counter',
              style: Theme.of(context).textTheme.display1,
            ),
          ],
        ),
      ),
      floatingActionButton: FloatingActionButton(
        onPressed: _incrementCounter,
        tooltip: 'Increment',
        child: Icon(Icons.add),
      ),
    );
  }

  @override
  void didChangeDependencies() {
    super.didChangeDependencies();
    print('didChangeDependencies');
  }

  @override
  void reassemble() {
    super.reassemble();    // 热重载回调，release模式下不起作用
    print('reassemble');
  }

  @override
  void didUpdateWidget(MyHomePage oldWidget) {
    super.didUpdateWidget(oldWidget);
    print('didUpdateWidget');
  }

  @override
```

```
  void deactivate() {
    super.deactivate();
    print('deactivate');
  }

  @override
  void dispose() {
    super.dispose();
    print('dispose');
  }
}
```

然后，我们创建一个路由来实现对界面的调整（关于路由的内容在后续章节会介绍，读者先不必纠结），代码如下所示：

```
import 'package:flutter/material.dart';

class NextPageWidget extends StatelessWidget {
  @override
  Widget build(BuildContext context) {
    return Scaffold(
      appBar: AppBar(
        title: Text('第2页'),
      ),
      body: Center(
        child: Text('当前是第2页'),
      ),
    );
  }
}
```

创建完这两个界面之后，我们就可以让项目"跑"（运行）起来了。在初次运行时，可以发现，在控制台依次打印了以下语句。

首次启动界面执行：initState→didChangeDependencies→build。

然后，我们点击一下热重载按钮（类似闪电的图标）。

依次执行：reassemble→didUpdateWidget→build。

注意：reassemble 只在 debug 模式下生效。

当界面发生跳转时，依次执行：deactivate→didChangeDependencies→build。

根据上面在代码中的执行顺序，整个流程可以用时序图表示，如图 3.35 所示。

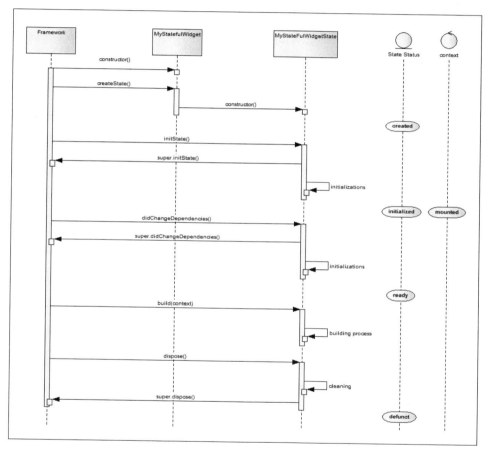

图 3.35

这里解释一下 State 生命周期里几个重要的方法。

1. initState

initState 是 State 生命周期里第一个被执行的方法，在实际项目中，当我们需要追加一些初始化方法的时候可以用其执行，比如 animations、controllers。如果要重写这个方法，就需要在这个方法里加上 super.initState()。在 initState 里，Framework 层还未把 Context 和 State 关联到一起，因此还不能访问 Context。另外，初始化方法在生命周期里只会被执行一次。

2．didChangeDependencies

该方法在执行完 initState 之后被执行，这个时候可以访问 Context 了。如果 Widget 使用了 InheritedWidget 的数据，并且在 InheritedWidget 的数据发生变化时，Flutter Framework 层就会触发 didChangeDependencies 的回调，关于 InheritedWidget 的知识在本章后面的小节会介绍。

3．build

build 方法在执行完 didChangeDependencies 和 didUpdateWidget 之后被执行，每当调用 setState((){…})方法时都会执行它。

4．dispose

dispose 方法在组件被销毁时调用，一般在执行 initState 初始化的动作时，dispose 都有对应的销毁动作。

3.4.8 Widget 的唯一身份标识：key

在 Flutter 里，每一个 Widget 都具有唯一标识，并且这个唯一标识是在 Flutter Framework 层创建和渲染时生成的，它就是 key。如果 key 作为参数被传入 Widget 里面，则会根据指定的名字生成 key。

在有些场景下，你需要保存 key，并且通过 key 访问该 Widget。这时，可以通过 GlobalKey、LocalKey、UniqueKey 或 ObjectKey 进行保存。例如，通过 GlobalKey 保存 key 并且在整个应用程序中共享，对应的代码如下所示：

```
GlobalKey myKey = new GlobalKey();
...
@override
Widget build(BuildContext context){
    return new MyWidget(
```

```
        key: myKey
    );
}
```

3.4.9 InheritedWidget

InheritedWidget 是一个比较特殊的组件，它被定义为父节点。被 InheritedWidget 暴露出来的数据，可以高效地在 Widget 树中从上往下传递和共享，并支持跨级数据传递。用一张图表示 InheritedWidget，如图 3.36 所示。

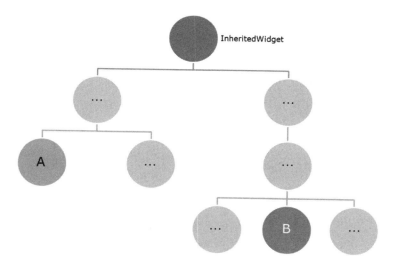

图 3.36

我们做一个购物车的例子，来体现 InheritedWidget 的作用，代码如下所示：

```
import 'package:flutter/material.dart';

class Item {
  String reference;

  Item(this.reference);
```

```dart
  }

  class _MyInherited extends InheritedWidget {
    _MyInherited({
      Key key,
      @required Widget child,
      @required this.data,
    }) : super(key: key, child: child);

    final MyInheritedWidgetState data;

    @override
    bool updateShouldNotify(_MyInherited oldWidget) {
      return true;
    }
  }

  class MyInheritedWidget extends StatefulWidget {
    MyInheritedWidget({
      Key key,
      this.child,
    }) : super(key: key);

    final Widget child;

    @override
    MyInheritedWidgetState createState() => new MyInheritedWidgetState();

    static MyInheritedWidgetState of(
        [BuildContext context, bool rebuild = true]) {
      return (rebuild
              ? context.inheritFromWidgetOfExactType(_MyInherited) as _MyInherited
              : context.ancestorWidgetOfExactType(_MyInherited) as _MyInherited)
          .data;
```

```dart
    }
  }

  class MyInheritedWidgetState extends State<MyInheritedWidget> {
    List<Item> _items = <Item>[];

    int get itemsCount => _items.length;

    void addItem(String reference) {
      setState(() {
        _items.add(new Item(reference));
      });
    }

    @override
    Widget build(BuildContext context) {
      return new _MyInherited(
        data: this,
        child: widget.child,
      );
    }
  }

  class MyTree extends StatefulWidget {
    @override
    _MyTreeState createState() => new _MyTreeState();
  }

  class _MyTreeState extends State<MyTree> {
    @override
    Widget build(BuildContext context) {
      return MyInheritedWidget(
        child: new Scaffold(
          appBar: new AppBar(
            title: new Text('Title'),
          ),
          body: new Column(
```

```
            children: <Widget>[
              new WidgetA(),
              new Container(
                child: new Row(
                  children: <Widget>[
                    new Icon(Icons.shopping_cart),
                    new WidgetB(),
                    new WidgetC(),
                  ],
                ),
              ),
            ],
          ),
        ),
      );
    }
  }

  class WidgetA extends StatelessWidget {
    @override
    Widget build(BuildContext context) {
      final MyInheritedWidgetState state = MyInheritedWidget.of(context, false);   // 防止 WidgetA 被 rebuild
      return new Container(
        child: new RaisedButton(
          child: new Text('Add Item'),
          onPressed: () {
            state.addItem('new item');
          },
        ),
      );
    }
  }

  class WidgetB extends StatelessWidget {
    @override
```

```
  Widget build(BuildContext context) {
    final MyInheritedWidgetState state = MyInheritedWidget.of
(context);
    return new Text('${state.itemsCount}');
  }
}

class WidgetC extends StatelessWidget {
  @override
  Widget build(BuildContext context) {
    return new Text('I am Widget C');
  }
}
```

在上面的例子中，_MyInherited 是 InheritedWidget 的实例，每当我们点击 "add Item" 添加数据时，_MyInherited 就会重新被创建。当 MyInheritedWidgetStated 对外暴露了 itemsCount 的 get 方法和 addItem 方法时，MyInheritedWidgetState 就会重新被创建。

完整的代码见 chapter3/flutter_widgets/lib/myinherited_widget.dart。

3.5 包管理

一个完整的应用或多或少会使用一些第三方包（包是为应用服务的）。在 Android 中，对包进行管理的是 gradle，而 iOS 用 Cocoapods 对包进行管理。在 Flutter 中，则通过 pubspec.yaml 配置文件对包进行管理。如图 3.37 所示，列出了一个简单的 pubspec.yaml 配置文件的情况。

我们可以在 dependencies 和 dev_dependencies 中加入所需要的依赖包，这些包都是官方或第三方开发者上传的。我们可以在 Pub 官方网站（https://pub.dartlang.org/）上搜到一些 Flutter 的第三方包，输入版本号之后，在项目的根目录里输入命令 "flutter packages get"，把包下载下来即可使用，或者在 IDE 里同步所需要的依赖包。

```
name: flutter_pingan
description: A new Flutter project.

# The following defines the version and build number for your application.
# A version number is three numbers separated by dots, like 1.2.43
# followed by an optional build number separated by a +.
# Both the version and the builder number may be overridden in flutter
# build by specifying --build-name and --build-number, respectively.
# Read more about versioning at semver.org.
version: 1.0.0+1

environment:
  sdk: ">=2.0.0-dev.68.0 <3.0.0"

dependencies:
  flutter:
    sdk: flutter
  json_annotation: ^2.0.0
  http: ^0.12.0

  # The following adds the Cupertino Icons font to your application.
  # Use with the CupertinoIcons class for iOS style icons.
  cupertino_icons: ^0.1.2

dev_dependencies:
  flutter_test:
    sdk: flutter
  build_runner: ^1.0.0
  json_serializable: ^2.0.0

# For information on the generic Dart part of this file, see the
# following page: https://www.dartlang.org/tools/pub/pubspec

# The following section is specific to Flutter.
flutter:

  # The following line ensures that the Material Icons font is
  # included with your application, so that you can use the icons in
  # the material Icons class.
  uses-material-design: true

  # To add assets to your application, add an assets section, like this:
  # assets:
```

图 3.37

我们可以自行开发一些插件上传到 Pub 上托管。pubspec.yaml 配置文件的配置项参数如表 3.10 所示。

表 3.10

参数	取值
name	表示应用或包的名称，项目里 import 的包名和这里设定的值一致
description	应用或包的描述信息
version	应用或包的当前版本号
dependencies	应用或包依赖的插件或其他第三方包
dev_dependencies	开发环境下依赖的包
flutter	一些配置项，例如 fonts 和 assets

3.6 常用代码段效果

本节会结合之前介绍的一些组件的使用方式，给出一些在实际项目中使用的案例。这些案例可以在构建项目时直接使用，能够快速地帮助大家创建和初始化一些常用的布局。具体的代码就不放在书中了，大家可以对应随书源码的地址自行下载使用。

3.6.1 案例一：侧滑效果

侧滑应用在 App 里非常普遍，它是通过 Scaffold 里的 Drawer 创建的，效果如图 3.38 所示。

图 3.38

完整的代码见 chapter3/flutter_slidemenu。

3.6.2 案例二：登录界面

不管什么样的 App，一般都会有登录界面，我们来看一下其效果，如图 3.39 所示。

图 3.39

读者可以根据实际情况，直接通过一些简单的修改就可以在项目里做出这种效果。

完整的代码见 chapter3/flutter_login。

3.6.3 案例三：轮播图效果

在 Flutter 里实现轮播图效果是比较容易的，只需要 PageView 就可以实现。由于轮播图在很多 App 里都有，这里就省略了示意图，读者可以直接下载代码使用。

完整的代码见 chapter3/flutter_pageview。

3.6.4 案例四：图片浏览器的相册效果

图片浏览器的相册效果在 App 里也是常见的，这里结合 CustomScrollView 和其中的 SliverGrid 来实现，效果如图 3.40 所示。

图 3.40

完整的代码见 chapter3/flutter_pictures。

3.6.5 案例五：全局主题设置

本例抽出了主题设置变量，变成了基本配置，方便配置全局主题。此处没有示意图，可以直接下载代码使用。

完整的代码见 chapter3/flutter_theme。

本章小结

笔者把组件分为三种：Basic-widget、Single-child 和 Multi-child。由于 Flutter 里的组件还是比较多的，为了便于记忆，读者可以根据自己的项目和实践心得，按照不同的功能做更好的划分。

写过前端的朋友一定知道，在 HTML 语言里有多种标签，其代表了各种"语义化"思想。类似的，在 Flutter 里面也希望大家能彻底掌握好每种组件的实际使用场景。每一种组件就是一种兵器，合理地运用好它们，便能在项目中更好地发挥作用，希望读者可以多加练习。

由于篇幅限制，本章只挑选一些重点的核心组件来讲解，要想查看更多的组件使用方法，请查看官方网站。在后续章节中，当碰到一些未在本章出现的布局和组件时，笔者也会说明其属性和使用场景。

第 4 章 事件处理

本章将为大家讲解在 Flutter 中如何实现一些跟移动端相关的事件处理机制,包括原始指针事件、GestureDetector(手势识别),以及事件的原理和分发机制。

4.1 原始指针事件

在移动设备上,H5(HTML5.0,超文本 5.0)开发和 Native 开发,其原始指针事件都是 Pointer Event。本书的原始指针事件指的是完成一次触摸的完整事件(手指按下—手指移动—手指抬起)。

需要指出的是,在手指按下之后,Flutter Framework 层是通过命中测试(Hit Test)获取到当前手指触摸的操作区域的,然后找到对应的 Widget。

4.1.1 基本用法

原始指针事件的常用回调方法如表 4.1 所示。

表 4.1

属性	取值
onPointerDown	手指按下回调
onPointerMove	手指移动回调
onPointerUp	手指抬起回调
onPointerCancel	触摸事件取消回调

具体的用法是把 Listener 包裹在需要监听的组件外来实现的。例如，下面的代码段就是比较基础的使用方式：

```dart
import 'package:flutter/gestures.dart';
import 'package:flutter/material.dart';

class ListenerPage extends StatefulWidget {
  @override
  _ListenerPageState createState() => new _ListenerPageState();
}

class _ListenerPageState extends State<ListenerPage> {
  String _opName = "未检测到操作";

  @override
  Widget build(BuildContext context) {
    return Scaffold(
      appBar: AppBar(
        title: Text("Listener 原始指针事件"),
      ),
      body: Center(
        child: Listener(
          child: Container(
            alignment: Alignment.center,
            color: Colors.blue,
            width: 240.0,
            height: 120.0,
            child: Text(
              _opName,
              style: TextStyle(color: Colors.white),
            ),
```

```
        ),
        onPointerDown: (PointerDownEvent event) =>
            _showEventText('onPointerDown'),
        onPointerMove: (PointerMoveEvent event) =>
            _showEventText('onPointerMove'),
        onPointerUp: (PointerUpEvent event) => _showEventText
('onPointerUp'),
        onPointerCancel: (PointerCancelEvent event) =>
            _showEventText('onPointerCancel'),
      ),
    ),
  );
}

void _showEventText(String text) {
  setState(() {
    _opName = text;
  });
  print(_opName);
}
}
```

运行之后可以看到控制台的输出，其顺序就是前面说的从手指按下到手指抬起的完整操作过程。在代码里，通过回调出来的参数我们可以看到有 PointerDownEvent、PointerMoveEvent、PointerUpEvent、PointerCancelEvent，其实都是 PointerEvent 的子类。PointerEvent 的构造器代码如下所示：

```
const PointerEvent({
  this.timeStamp = Duration.zero,
  this.pointer = 0,
  this.kind = PointerDeviceKind.touch,
  this.device = 0,
  this.position = Offset.zero,
  this.delta = Offset.zero,
  this.buttons = 0,
  this.down = false,
  this.obscured = false,
  this.pressure = 1.0,
```

```
    this.pressureMin = 1.0,
    this.pressureMax = 1.0,
    this.distance = 0.0,
    this.distanceMax = 0.0,
    this.size = 0.0,
    this.radiusMajor = 0.0,
    this.radiusMinor = 0.0,
    this.radiusMin = 0.0,
    this.radiusMax = 0.0,
    this.orientation = 0.0,
    this.tilt = 0.0,
    this.platformData = 0,
    this.synthesized = false,
});
```

通过上面列出的参数，我们可以获取一些与手势相关的参数值。这些都是系统告诉我们的，读者感兴趣的话可以把想要了解的值打印出来。需要具体说明并且要掌握的参数如下所示。

（1）position：相对于全局坐标的偏移。

（2）delta：两次指针移动事件的距离。

（3）orientation：指针移动方向。

完整的代码见 chapter4/flutter_event/lib/listener_page.dart。

4.1.2 忽略事件

有的时候，我们不需要响应 PointerEvent，这个时候可以使用 IgnorePointer 和 AbsorbPointer。但是，两者是有区别的，我们分别说明一下。

IgnorePointer：此节点和其子节点都将忽略点击事件，用 ignoring 参数区分是否忽略。

AbsorbPointer：这个控件本身是能够响应点击事件的，它做的事情是阻止事件传播到它的子节点上去。

可见，IgnorePointer 与 AbsorbPointer 的区别是前者本身不再接收事件，而后者本身可以接收事件。

我们来看一个例子，代码如下所示：

```
import 'package:flutter/material.dart';

class PointerEventIgnorePage extends StatefulWidget {
  @override
  _PointerEventIgnorePageState createState() =>
      new _PointerEventIgnorePageState();
}

class _PointerEventIgnorePageState extends State
<PointerEventIgnorePage> {
  bool _ignore = true;

  @override
  Widget build(BuildContext context) {
    return Scaffold(
      appBar: AppBar(title: Text('忽略事件')),
      body: Container(
        alignment: AlignmentDirectional.center,
        child: Column(
          children: <Widget>[
            Switch(
              value: _ignore,
              onChanged: (value) {
                setState(() => _ignore = value);
              },
            ),
            GestureDetector(
              onTap: () => print('GestureDetector Clicked!'),
              child: IgnorePointer(
                ignoring: _ignore,
                child: RaisedButton(
                  onPressed: () => print('IgnorePointer Clicked!'),
```

```
                child: Text('IgnorePointer'),
              ),
            ),
          ),
          GestureDetector(
            onTap: () => print('GestureDetector Clicked!'),
            child: AbsorbPointer(
              absorbing: _ignore,
              child: RaisedButton(
                onPressed: () => print('AbsorbPointer Clicked!'),
                child: Text('AbsorbPointer'),
              ),
            ),
          ),
        ],
      ),
    ),
  );
 }
}
```

代码运行之后,结果如图 4.1 所示。

图 4.1

当我们开启忽略事件时,点击 AbsorbPointer 按钮,控制台就会打印,而点击 IgnorePointer 按钮之后控制台不会打印。

4.2 GestureDetector

有的时候，原始指针事件的一些操作并不能满足我们的使用，这个时候就需要更强大的手势识别组件，来支持缩放、双击、垂直、水平等补充手势，这个组件就是 GestureDetector（手势识别），这一节给大家介绍一下该组件的使用方法。

4.2.1 基本用法

GestureDetector 通常作为一个父 Widget 包裹在一个子 Widget 外面，它是组件的一种，并且我们可以通过 onTap 回调来实现其点击的效果，例如：

```
GestureDetector(
  onTap: () {
    print("tap");
  },
  child: Container(
    padding: EdgeInsets.all(12.0),
    decoration: BoxDecoration(
      color: Theme.of(context).buttonColor,
      borderRadius: BorderRadius.circular(8.0),
    ),
    child: Text('一个按钮'),
  ),
);
```

这是最简单、最基本的用法，只有一个点击事件。本例中在 onTap 时做了 print 操作，用户可以根据实际项目需求设置 onTap 事件触发的动作，比如路由切换或者弹出对话框。

注意，如果你希望能在点击时看见水波效果，则可以使用 InkWell 代替 GestureDetector，其他具有类似效果的组件还有 RaisedButton、FlatButton、CupertinoButton。

4.2.2 常用事件

上一节讲了 GestureDetector 的基本用法，同时也是最常用的用法。实际上我们与手机的交互行为远不止 onTap 这一个，通过 GestureDetector 管理手势，还支持如表 4.2 所示的手势。

表 4.2

属　　性	取值意义
onTapDown	当按下屏幕时触发
onTap	当与屏幕短暂地触碰时触发，最常用
onTapUp	当用户停止触碰屏幕时触发
onTapCancel	当用户触摸屏幕、但没有完成 Tap 事件时触发
onDoubleTap	当快速双击屏幕时触发
onLongPress	当长按屏幕时触发（与屏幕接触的时间超过 500ms）
onPanUpdate	当在屏幕上移动时触发
onVerticalDragDown	当手指触碰屏幕且准备往屏幕垂直方向移动时触发
onVerticalDragStart	当手指触碰屏幕且开始往屏幕垂直方向移动时触发
onVerticalDragUpdate	当手指触碰屏幕且开始往屏幕垂直方向移动并发生位移时触发
onVerticalDragEnd	当用户完成垂直方向触摸屏幕时触发
onVerticalDragCancel	当用户中断了 onVerticalDragDown 时触发
onHorizontalDragDown	当手指触碰屏幕且准备往屏幕水平方向移动时触发
onHorizontalDragStart	当手指触碰屏幕且开始往屏幕水平方向移动时触发
onHorizontalDragUpdate	当手指触碰屏幕且开始往屏幕水平方向移动并发生位移时触发
onHorizontalDragEnd	当用户完成水平方向触摸屏幕时触发
onHorizontalDragCancel	当用户中断了 onHorizontalDragDown 时触发
onPanDown	当用户触摸屏幕时触发
onPanStart	当用户触摸屏幕并开始移动时触发
onPanUpdate	当用户触摸屏幕并产生移动时触发
onPanEnd	当用户完成触摸屏幕时触发
onScaleStart	当用户触摸屏幕并开始缩放时触发
onScaleUpdate	当用户触摸屏幕并产生缩放时触发
onScaleEnd	当用户完成缩放时触发

以一个例子来说明，代码如下所示：

```
import 'package:flutter/material.dart';

class GestureDetectorPage extends StatefulWidget {
  @override
  _GestureDetectorState createState() => new _GestureDetectorState();
}

class _GestureDetectorState extends State<GestureDetectorPage> {
  String _opName = "未检测到操作";

  @override
  Widget build(BuildContext context) {
    return Scaffold(
      appBar: AppBar(
        title: Text("GestureDetector手势识别"),
      ),
      body: Center(
        child: GestureDetector(
          child: Container(
            alignment: Alignment.center,
            color: Colors.blue,
            width: 240.0,
            height: 120.0,
            child: Text(
              _opName,
              style: TextStyle(color: Colors.white),
            ),
          ),
          onTap: () => _showEventText("Tap"),
          onTapUp: (e) => _showEventText("TapUp"),
          onTapDown: (e) => _showEventText("TapDown"),
          onTapCancel: () => _showEventText("TapCancel"),
          onDoubleTap: () => _showEventText("DoubleTap"),
          onLongPress: () => _showEventText("LongPress"),
          onVerticalDragDown: (e) => _showEventText("onVerticalDragDown"),
          onVerticalDragStart: (e) => _showEventText
```

```
("onVerticalDragStart"),
          onVerticalDragUpdate: (e) => _showEventText
("onVerticalDragUpdate"),
          onVerticalDragEnd: (e) => _showEventText
("onVerticalDragEnd"),
          onVerticalDragCancel: () => _showEventText
("onVerticalDragCancel"),
          onHorizontalDragDown: (e) => _showEventText
("onHorizontalDragDown"),
          onHorizontalDragStart: (e) => _showEventText
("onHorizontalDragStart"),
          onHorizontalDragUpdate: (e) => _showEventText
("onHorizontalDragUpdate"),
          onHorizontalDragEnd: (e) => _showEventText
("onHorizontalDragEnd"),
          onHorizontalDragCancel: () => _showEventText
("onHorizontalDragCancel"),
    //          onPanDown: (e) => _showEventText("onPanDown"),
    //          onPanStart: (e) => _showEventText("onPanStart"),
    //          onPanUpdate: (e) => _showEventText("onPanUpdate"),
    //          onPanEnd: (e) => _showEventText("onPanEnd"),
    //          onScaleStart: (e) => _showEventText("onScaleStart"),
    //          onScaleUpdate: (e) => _showEventText("onScaleUpdate"),
    //          onScaleEnd: (e) => _showEventText("onScaleEnd"),
        ),
      ),
    );
  }

  void _showEventText(String text) {
    setState(() {
      _opName = text;
    });
    print(_opName);
  }
}
```

我们在手机屏幕上定义一个矩形区域，当以各种方式通过手指与屏幕交互时都会触发对应的回调事件。一次完整的手势触摸屏幕的过程应该是手指按下、移动（可能用户没移动）、抬起的完整过程，以垂直方向回调事件序列为例，执行顺序如下所示：

onVerticalDragDown → onVerticalDragStart → onVerticalDragUpdate → onVerticalDragEnd

建议读者把本例多运行几次，体验一下，根据前面列出的属性取值来理解，这样会有更为直观的感受。

注意：onVerticalUpdate、onHorizontalUpdate、onPadUpdate 这些事件不能同时存在，否则会报错，报错提示信息如图 4.2 所示。

```
flutter: Simultaneously having a vertical drag gesture recognizer, a horizontal drag gesture recognizer, and
flutter: a pan gesture recognizer will result in the pan gesture recognizer being ignored, since the other
flutter: two will catch all drags.
```

图 4.2

出现这种报错和 Flutter 事件处理的竞争机制有关。

另外，onPanUpdate 和 onScaleUpdate 也不能同时存在，这是因为在 Gesture 识别器里，Scale 操作是 Pan 操作的超集。

完整的代码见 chapter4/flutter_event/lib/gesture_detector_page.dart。

4.2.3 拖曳和缩放效果

上一节我们学习了 onPanUpdate 和 onScaleUpdate 这类事件。现在，我们通过一些简单的例子来具体说明一下。比如通过手势改变某一个控件的位置并实现拖曳，代码如下所示：

```
import 'package:flutter/material.dart';

class DragPage extends StatefulWidget {
  @override
  _DragState createState() => new _DragState();
```

```dart
    }

    class _DragState extends State<DragPage> with SingleTickerProviderStateMixin {
      double _top = 0.0;
      double _left = 0.0;
      double _size = 100.0;

      @override
      Widget build(BuildContext context) {
        return Scaffold(
          appBar: AppBar(
            title: Text("拖曳控件"),
          ),
          body: Stack(
            children: <Widget>[
              Positioned(
                top: _top,
                left: _left,
                child: GestureDetector(
                  child: FlutterLogo(
                    size: _size,
                  ),
                  onPanUpdate: (e) {
                    setState(() {
                      _left += e.delta.dx;
                      _top += e.delta.dy;
                    });
                  },
                ),
              )
            ],
          ),
        );
      }
    }
```

从代码中我们可以看出，当执行 onPanUpdate 回调时，通过 setState 也改变了 Positioned 组件的位置，即在拖曳过程中不断地改变 let 和 top 的值。这样看起来就实现了拖曳，我们也可以把 onPanUpdate 改成 onVerticalDragUpdate 或 onHorizontalDragUpdate，变成单一方向的拖曳效果。

完整的代码见 chapter4/flutter_event/lib/drag_page.dart。

我们再来看看，如何用 onScaleUpdate 手势识别来实现缩放效果。其实只需要稍微改一下代码，具体代码如下所示：

```dart
import 'package:flutter/material.dart';

class ScalePage extends StatefulWidget {
  @override
  _ScaleState createState() => new _ScaleState();
}

class _ScaleState extends State<ScalePage> {
  double _size = 100.0;

  @override
  Widget build(BuildContext context) {
    return Scaffold(
      appBar: AppBar(
        title: Text("缩放识别"),
      ),
      body: Center(
        child: GestureDetector(
          child: Container(
            alignment: Alignment.center,
            child: FlutterLogo(size: _size,),
          ),
          onScaleUpdate: (ScaleUpdateDetails e) {
            setState(() {
              _size = 300 * e.scale.clamp(.5, 10.0);
            });
          },
        ),
```

```
      ),
    );
  }
}
```

完整的代码见 chapter4/flutter_event/lib/scale_page.dart。

4.2.4 事件竞争与手势冲突

其实在讲 GestureDetector 常用事件时，已经提到过水平和垂直事件的监听。但是运行代码时我们发现，应用要么只执行 onVerticalUpdate 回调，要么只执行 onHorizontalUpdate 回调，这是什么原因呢？这个时候事件竞争机制的处理就登场了。

Flutter 加入了手势竞技场（Gesture Arena）的概念。在给同一个组件同时加入水平和垂直回调时，若用户将指针水平移动超过一定的逻辑像素，则水平识别器将声明胜利，并且手势将被解释为水平拖曳。同理，用户垂直移动超过一定的逻辑像素，则垂直识别器将宣布胜利。我们来看一个例子，代码如下所示：

```
import 'package:flutter/gestures.dart';
import 'package:flutter/material.dart';

class BothDirectionPage extends StatefulWidget {
  @override
  _BothDirectionState createState() => new _BothDirectionState();
}

class _BothDirectionState extends State<BothDirectionPage> {
  double _top = 0.0;
  double _left = 0.0;

  @override
  Widget build(BuildContext context) {
    return Scaffold(
      appBar: AppBar(
```

```
        title: Text("事件竞争机制"),
    ),
    body: Stack(
      children: <Widget>[
        Positioned(
          top: _top,
          left: _left,
          child: GestureDetector(
            child: FlutterLogo(
              size: 100,
            ),
            onVerticalDragUpdate: (DragUpdateDetails e) {
              setState(() {
                _top += e.delta.dy;
              });
              print("垂直胜出");
            },
            onHorizontalDragUpdate: (DragUpdateDetails e) {
              setState(() {
                _left += e.delta.dx;
              });
              print("水平胜出");
            },
            onTapDown: (e) {
              print("按下");
            },
            onTapUp: (e) {
              print("松开");
            },
            onHorizontalDragEnd: (e) {
              print("水平移动结束");
            },
            onVerticalDragEnd: (e) {
              print("垂直移动结束");
            },
          ),
        )
      ],
```

```
      ),
    );
  }
}
```

在以上代码中，还有几个回调方法 onTapDown、onTapUp、onVerticalDragEnd。如果我们只是简单地触碰一下屏幕然后松开，则执行 onTapDown→onTapUp 回调。而这个时候，如果我们进行了垂直方向的位移，即触发了 onVerticalDragUpdate 回调，则松开手后 onTapUp 不会被执行，取而代之的是 onVerticalDragEnd 回调方法被执行。这种情况就是手势冲突。

完整的代码见 chapter4/flutter_event/lib/both_direction_page.dart。

4.2.5 手势识别器

手势识别器即 GestureRecognizer，是 GestureDetector 的一部分，而 GestureDetector 可以用一个或多个 GestureRecognizer 识别各种手势。GestureRecognizer 是一个被抽象出来的类，该类有多种实现方式的子类，比如 LongPressGestureRecognizer、TapGestureRecognizer，这是 Flutter 为我们实现的。我们以一个查单词的 App 为例，代码如下所示：

```
import 'package:flutter/gestures.dart';
import 'package:flutter/material.dart';

class GestureRecognizerPage extends StatefulWidget {
  @override
  _GestureRecognizerState createState() => new
_GestureRecognizerState();
}

class _GestureRecognizerState extends State
<GestureRecognizerPage> {
    TapGestureRecognizer _tapGestureRecognizer = new
TapGestureRecognizer();
    final GlobalKey<ScaffoldState> _scaffoldKey = new
GlobalKey<ScaffoldState>();
```

```
@override
void dispose() {
  _tapGestureRecognizer.dispose();
  super.dispose();
}

@override
Widget build(BuildContext context) {
  return Scaffold(
    key: _scaffoldKey,
    appBar: AppBar(
      title: Text("GestureRecognizer"),
    ),
    body: Padding(
      padding: EdgeInsets.all(10.0),
      child: Column(
        children: <Widget>[
          Text.rich(
            TextSpan(
              children: [
                TextSpan(
                  text: "Room is not ",
                  style: TextStyle(fontSize: 25.0),
                ),
                TextSpan(
                  text: "built",
                  style: TextStyle(
                      fontSize: 25.0,
                      color: Colors.blue,
                      fontWeight: FontWeight.bold),
                  recognizer: _tapGestureRecognizer
                    ..onTap = () {
                      showInSnackBar("built: 建造");
                    },
                ),
                TextSpan(
                  text: " in one day.",
                  style: TextStyle(fontSize: 25.0),
                ),
              ],
            ),
          ),
```

```
        ],
      ),
    ),
  );
}

void showInSnackBar(String value) {
  _scaffoldKey.currentState.showSnackBar(
    new SnackBar(
      content: new Text(
        value,
        style: TextStyle(fontSize: 25.0),
      ),
    ),
  );
}
```

运行效果如图 4.3 所示。

图 4.3

在本例中我们点击句子里高亮标记的单词，就会通过 SnackBar 弹出单词的中文意思。TextSpan 不是 Widget，所以只接受一个 Recognizer。这里运用了 TapGestureRecognizer 识别器，用它作用于 TextSpan。于是，我们就通过 TapGestureRecognizer 实现了该功能。值得注意的是，我们需要在 State 生命周期的 dispose 方法中，调用 TapGestureRecognizer 的 dispose 方法来释放资源。

完整的代码见 chapter4/flutter_event/lib/gesture_recognizer_page.dart。

4.3 事件原理与分发机制

前面我们已经了解了 Gesture、Listener 二者与手势相关的事件。在 Flutter Framework 层中，手势实现的相关原理是什么呢？本节将仔细讲解。

首先，我们看一下 Flutter 框架的入口方法，代码如下所示：

```
void main() => runApp(new MyApp());

void runApp(Widget app) {
  WidgetsFlutterBinding.ensureInitialized()
    ..attachRootWidget(app)
    ..scheduleWarmUpFrame();
}
```

这些代码是我们在初始化 Flutter 工程时就有的。在 Android Studio 里，我们进入 WidgetsFlutterBinding.ensureInitialized()方法去查看它具体都做了什么，代码如下所示：

```
class WidgetsFlutterBinding extends BindingBase with
GestureBinding, ServicesBinding, SchedulerBinding, PaintingBinding,
SemanticsBinding, RendererBinding, WidgetsBinding {
  // ...
}
```

我们可以看到，Flutter 用到了 Mixin 的语言特性，依次执行了 BindingBase、GestureBinding、ServicesBinding、SchedulerBinding、

PaintingBinding、SemanticsBinding、RendererBinding、WidgetsBinding 的 initInstances。在手势里面，与之关联的就是 GestureBinding 和 RendererBinding。

在 GestureBinding 里，初始化方法如下所示：

```
@override
void initInstances() {
  super.initInstances();
  _instance = this;
  ui.window.onPointerDataPacket = _handlePointerDataPacket;
}
```

这里可以看到，在初始化中，该方法把_handlePointerDataPacket 传递给了 ui.window.onPointerDataPacket，而_handlePointerDataPacket 的代码如下所示：

```
void _handlePointerDataPacket(ui.PointerDataPacket packet) {
  _pendingPointerEvents.addAll(PointerEventConverter.expand(packet.data, ui.window.devicePixelRatio));
    if (!locked)
      _flushPointerEventQueue();
  }
```

可以看出，在获取参数 ui.PointerDataPacket 之后，从里面取出了 data 数据包，然后通过 PointerEventConverter 的 expand 静态方法，即 PointerEventConverter.expand 方法扩展添加到 Framework 层的 point events 中并返回 Iterable 可迭代对象，最后添加到_pendingPointerEvents 中。

这个时候，通过尝试调用_flushPointerEventQueue 方法依次处理每一个 pointer event。这样设计是因为前面的 pointer 事件需要先被处理，所以采用了 queue 队列的数据结构，代码如下所示：

```
void _flushPointerEventQueue() {
  assert(!locked);
  while (_pendingPointerEvents.isNotEmpty)
    _handlePointerEvent(_pendingPointerEvents.removeFirst());
}
```

_pendingPointerEvents 是 Queue 的实例，如果_pendingPointerEvents 队列不为空，则执行_handlePointerEvent 方法。该方法也是手势处理的核心，代码段如下所示：

```
void _handlePointerEvent(PointerEvent event) {
  assert(!locked);
  HitTestResult result;
  if (event is PointerDownEvent) {
    assert(!_hitTests.containsKey(event.pointer));
    result = HitTestResult();
    hitTest(result, event.position);
    _hitTests[event.pointer] = result;
    assert(() {
      if (debugPrintHitTestResults)
        debugPrint('$event: $result');
      return true;
    }());
  } else if (event is PointerUpEvent || event is PointerCancelEvent) {
    result = _hitTests.remove(event.pointer);
  } else if (event.down) {
    result = _hitTests[event.pointer];
  } else {
    return; // We currently ignore add, remove, and hover move events.
  }
  if (result != null)
    dispatchEvent(event, result);
}
```

在上面的代码中，如果 event 是在 PointerDownEvent 的情况下，则会初始化 HitTestResult，并把 HitTestResult 作为参数传入 hitTest，代码如下所示：

```
@override // from HitTestable
void hitTest(HitTestResult result, Offset position) {
  result.add(HitTestEntry(this));
}
```

通过 hitTest 检查 pointer event 涉及了哪些 Widget，并判断 pointer event 的 position 命中了哪些 view。然后把这些 Widget 添加到 HitTestResult 的 path 中，pointer event 会依次在最里层的 Widget 树中按照路径 path 进行查找。经过 path 进行事件分发，到达与之相关的所有 Widget，这和前端的冒泡事件有点相似。

具体而言，可以通过 dispatchEvent 方法实现，代码如下所示：

```
@override // from HitTestDispatcher
void dispatchEvent(PointerEvent event, HitTestResult result) {
  assert(!locked);
  assert(result != null);
  for (HitTestEntry entry in result.path) {
    try {
      entry.target.handleEvent(event, entry);
    } catch (exception, stack) {
      FlutterError.reportError(FlutterErrorDetailsForPointerEventDispatcher(
        exception: exception,
        stack: stack,
        library: 'gesture library',
        context: 'while dispatching a pointer event',
        event: event,
        hitTestEntry: entry,
        informationCollector: (StringBuffer information) {
          information.writeln('Event:');
          information.writeln(' $event');
          information.writeln('Target:');
          information.write(' ${entry.target}');
        }
      ));
    }
  }
}
```

在 dispatchEvent 方法里，通过遍历 result.path 里的 HitTestEntry 作为参数，通过调用 entry.target.handleEvent 进行具体的事件处理，handleEvent 的具体逻辑如下所示：

```
@override // from HitTestTarget
void handleEvent(PointerEvent event, HitTestEntry entry) {
  pointerRouter.route(event);
  if (event is PointerDownEvent) {
    gestureArena.close(event.pointer);
  } else if (event is PointerUpEvent) {
    gestureArena.sweep(event.pointer);
  }
}
```

在上面的代码中,"pointerRouter.route(event);"对应的源码如下所示:

```
void route(PointerEvent event) {
  final LinkedHashSet<PointerRoute> routes = _routeMap[event.pointer];
  final List<PointerRoute> globalRoutes = List<PointerRoute>.from(_globalRoutes);
  if (routes != null) {
    for (PointerRoute route in List<PointerRoute>.from(routes)) {
      if (routes.contains(route))
        _dispatch(event, route);
    }
  }
  for (PointerRoute route in globalRoutes) {
    if (_globalRoutes.contains(route))
      _dispatch(event, route);
  }
}
```

routes 属于 LinkedHashSet,会被有序地在 globalRoutes 里遍历并分发。我们在自定义手势识别类时,会对其中的 handleEvent 的具体逻辑进行 override,并在里面根据我们的自定义手势逻辑执行相应的动作。手势处理完成后,如果是 PointerDownEvent,则关闭 gestureArena(手势竞技场);如果是 PointerUpEvent,则清理 gestureArena。

4.4 事件通知

Notification 是"通知"的意思，这和 Android 中不一样。在 Flutter 里，Notification 会沿着当前的 context 节点从下往上传递，所有父节点都可以通过 NotificationListener 来监听通知，这种由子向父的传递方式为"通知冒泡"。Flutter 里的 Notification 有点类似前端的"冒泡事件"。我们可以定义通知类，并继承至 Notification，而父 Widget 使用 NotificationListener 进行监听并捕获通知。常用的 NotificationListener 有 LayoutChangeNotification、SizeChangedLayoutNotifier、ScrollNotification 等。比如，我们将 ListView 加上 NotificationListener，并通过 NotificationListener 里的 onNotification 回调方法来判断滚动状态，具体代码如下所示：

```
import 'package:flutter/material.dart';

class ScrollStatusPage extends StatefulWidget {
  @override
  _ScrollStatusState createState() => new _ScrollStatusState();
}

class _ScrollStatusState extends State<ScrollStatusPage> {
  String message = "";

  _onStartScroll(ScrollMetrics metrics) {
    setState(() {
      message = "Scroll Start";
    });
  }

  _onUpdateScroll(ScrollMetrics metrics) {
    print(metrics.pixels);
    setState(() {
      message = "Scroll Update";
    });
  }
```

```dart
  _onEndScroll(ScrollMetrics metrics) {
    setState(() {
      message = "Scroll End";
    });
  }

  @override
  Widget build(BuildContext context) {
    return Scaffold(
      appBar: AppBar(
        title: Text("NotificationListener"),
      ),
      body: Column(
        children: <Widget>[
          Container(
            height: 50.0,
            color: Colors.green,
            child: Center(
              child: Text(message),
            ),
          ),
          Expanded(
            child: NotificationListener<ScrollNotification>(
              onNotification: (scrollNotification) {
                if (scrollNotification is ScrollStartNotification) {
                  _onStartScroll(scrollNotification.metrics);
                } else if (scrollNotification is ScrollUpdateNotification) {
                  _onUpdateScroll(scrollNotification.metrics);
                } else if (scrollNotification is ScrollEndNotification) {
                  _onEndScroll(scrollNotification.metrics);
                }
              },
              child: ListView.builder(
                itemCount: 30,
                itemBuilder: (context, index) {
                  return ListTile(title: Text("Index : $index"));
```

```
                    },
                ),
              ),
            ),
          ],
        ),
      );
    }
  }
```

在本例中，我们监听了 ListView 的滚动状态，分别是 Scroll Start、Scroll Update、Scroll End。当用户滚动时，就可以在屏幕上显示出滚动状态，完整的效果如图 4.4 所示。

图 4.4

完整的代码见 chapter4/flutter_event/lib/scroll_status_page.dart。

本章小结

通过本章的学习，相信大家对事件处理有了简单的了解。在学好事件处理之后，我们就能够结合一些 UI 来自定义并做出更好的 App 效果了，自下一章开始，我们将学习动画。

第 5 章

动　　画

在上一章节中，我们介绍了 Flutter 事件的处理机制和原理，本章将为大家讲解动画相关的内容。只要具备了手势处理和动画相关的知识，我们就能创造内容丰富的自定义控件了。

不管是 iOS 平台，还是 Android 平台，用户在使用 App 时都能看到一些动画效果。交互动画可以作为用户的操作向导，也可以让用户心情愉悦，没有动画的 App 可以说是枯燥无味的。本章开始，我们就来聊一聊动画在 Flutter 中是如何使用的。

5.1 动画原理及概述

提到动画原理，就不得不说"动画帧"。比如手机拍摄的视频，虽然看起来展现的是一个连续播放的视频，但其实视频本身都是由一幅幅静态的图片组成的。静态图片的连续展示就是一段视频。

这里有一个概念，即画面每秒传输帧数（Frames Per Second，FPS）。FPS 的值越高，动画就越流畅。目前大多数设备的屏幕刷新率为 60Hz，所以，在通常情况下 FPS 为 60fps 时效果最佳，也就是每帧消耗的时间为 16.67ms。在 Flutter 中，FPS 已经达到 60fps，所以动画的流畅性也已达到原生动画的效果。

动画的核心是 Animation 类，它可以判断当前的动画状态（比如开始、停止、移动、前进、反向），但是它不关心屏幕上显示的任何组件。Animation 是由 AnimationController 管理的，并通过 Listeners 和 StatusListeners 管理动画状态所发生的变化。在 Flutter 中，通过 Animation、AnimationController、Tween、Curve 对动画进行了抽象（模型化）。

5.1.1 Animation

在动画中使用最多的类型是 Animation。Animation 对象本身对手机屏幕是无感知的，而且，它对调用的 build 方法也是无感知的。它是一个抽象类，仅仅知道当前动画的插值和状态（completed 或 dismissed）。Animation 对象是一个在一段时间内，依次生成一个区间值的类，其输出值可以是线性的、非线性的，也可以是一个步进函数或者其他曲线函数。控制器可以决定 Animation 动画的运行方式：正向、反向以及在中间进行切换。Animation 也可以生成除 double 之外的其他值，比如 Animation<Color>或 Animation<Size>。

5.1.2 Animatable

Animatable 是控制动画类型的类。比如在平移动画中，我们关心的是 (x,y) 的值，那么这个时候就需要 Animatable 控制 (x,y) 的值的变化。在颜色动画（ColorTween）中，我们关心的是色值的变化，那么就需要 Animatable 控制色值的变化。在贝塞尔曲线运动中，我们关心的是路径是否按照贝塞尔方程式来生成 (x,y)，所以 Animatable 就要按照贝塞尔方程式的方式来改变 (x,y)。

5.1.3 AnimationController

AnimationController 即动画控制器，它负责在给定的时间段内，以线性的方式生成默认区间为(0.0,1.0)的数字。我们可以通过 AnimationController 来创建 Animation 对象，例如：

```
final AnimationController controller = AnimationController(
    duration: const Duration(seconds: 2), vsync: this);
```

以上 AnimationController 的创建派生自 Animation，它能告诉 Flutter 动画的控制器已经创建好对象了，并且处于准备状态。要真正让动画运作起来，则需要通过 AnimationController 的 forward()方法来启动动画。其中，数值的产生取决于屏幕的刷新情况，一般每秒产生 60 帧。数值生成之后，每个 Animation 对象都会通过 Listener 进行回调。例如，图 5.1 所示的控制台的 animation.value 的值，就是通过 Listener 监听得到的。

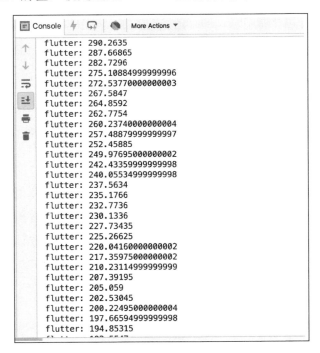

图 5.1

在动画的状态发生变化时，则可以调用 addStatusListener 来监听，代码如下所示：

```
animation = Tween(begin: 0.0, end: 300.0).animate(controller)
    ..addStatusListener((status) {
        if (status == AnimationStatus.completed) {
          controller.reverse();
        } else if (status == AnimationStatus.dismissed) {
          controller.forward();
        }
    });
```

在上述代码中，AnimationController 可以调用 reverse 方法，并且在需要的时候还可以调用 stop 方法。

当创建 AnimationController 时，需要传入 vsync 参数，这个参数接受的是 TickerProvidr 类型的对象，作用是阻止在屏幕锁屏时执行动画以避免不必要的资源浪费。

我们来看一个例子，代码如下所示：

```
// ...
class _LogoAppState extends State<LogoApp> with
SingleTickerProviderStateMixin {
  Animation<double> animation;
  AnimationController controller;

  initState() {
    super.initState();
    controller = AnimationController(
        duration: const Duration(milliseconds: 2000), vsync: this);
    // ...
    controller.forward();
  }
}
```

上述例子是将 SingleTickerProviderStateMixin 添加到 State 中，并设置了 vsync 的值。

5.1.4 Tween

Tween 即补间动画。

在通常情况下，AnimationController 的取值范围是(0.0,1.0)。但是，若我们需要不同的范围或类型的值时，则可以使用 Tween 来定义并生成不同的范围或类型的值。例如：

```
final Tween doubleTween = Tween<double>(begin: -200.0, end: 0.0);
```

从上述代码中可以看出，Tween 生成(-200.0,0.0)的值。

在 Tween 构造函数时，需要 begin 和 end 两个参数。Tween 的唯一作用就是定义从输入范围到输出范围的映射，输入范围一般为(0.0,1.0)，但这不是必需的，我们可以自定义。

Tween 继承自 Animatable，而不是 Animation。虽然 Animatable 与 Animation 相似，但它不是必须输出 double 值。比如，我们通过 ColorTween 实现两个颜色从开始（begin）到结束（end）的渐变，代码如下所示：

```
final Tween colorTween = ColorTween(begin: Colors.transparent, end: Colors.black54);
```

从上述代码中可以看出，我们定义了 ColorTween 颜色之间的取值。

虽然 Tween 不会存储任何状态,但它提供了 evaluate(Animation<double> animation)方法，并可以通过映射获取动画当前的值。Animation 当前的值可以通过 value 方法来获取。evaluate 还能执行其他处理，比如确保动画值分别为 0.0 和 1.0 时，返回开始和结束状态。若要使用 Tween 对象，需要调用 animate 方法，并传入一个控制器对象。

通过上面的学习，我们再看一个与 Tween 相关的例子，代码如下所示：

```
import 'package:flutter/material.dart';

class TweenAnimation extends StatefulWidget {
  TweenAnimation({Key key, this.title}) : super(key: key);
```

```
  final String title;

  @override
  _TweenAnimationState createState() => _TweenAnimationState();
}

class _TweenAnimationState extends State<TweenAnimation>
    with SingleTickerProviderStateMixin {
  Animation<double> animation;
  AnimationController controller;

  @override
  void initState() {
    super.initState();
    controller = AnimationController(
        duration: const Duration(milliseconds: 2000), vsync: this);
    animation = Tween(begin: 0.0, end: 300.0).animate(controller)
      ..addListener(() {
        setState(() {
          print(animation.value);
        });
      })
      ..addStatusListener((status) {
        // 实现动画循环播放
        if (status == AnimationStatus.completed) {
          controller.reverse();
        } else if (status == AnimationStatus.dismissed) {
          controller.forward();
        }
      });
    controller.forward();
  }

  dispose() {
    controller.dispose();
    super.dispose();
  }
```

```
  @override
  Widget build(BuildContext context) {
    return Scaffold(
      appBar: AppBar(
        title: Text("缩放动画"),
      ),
      body: Container(
        margin: EdgeInsets.symmetric(vertical: 10.0),
        height: animation.value,
        width: animation.value,
        child: FlutterLogo(),
      ));
  }
}
```

上述代码演示了图片放大再缩小,并不断循环的过程。通过 Status 可以判断 AnimationStatus.completed 和 AnimationStatus.dismissed 的两种状态。它们分别调用了 controller.reverse()方法和 controller.forward()方法。

注意:在上述代码中,出现了两个点 ".." 的语法,其是级联操作符,在第 2 章我们已经接触过,例如:

```
animation = tween.animate(controller)
    ..addListener(() {
      setState(() {
        // the animation object's value is the changed state
      });
    });
```

这种写法是执行了 animate()方法返回值之后再执行的方法,等价于下面的写法,代码如下所示:

```
animation = tween.animate(controller);
animation.addListener(() {
    setState(() {
      // the animation object's value is the changed state
    });
});
```

完整的代码见 chapter5/flutter_animations/lib/tween_animation.dart。

5.1.5　Tween.animate

如果要使用 Tween 对象，就需要调用其 animate 方法，并传入一个控制器对象。例如，实现在 500ms 内生成从 0 到 255 的整数值，代码如下所示：

```
AnimationController controller = AnimationController(
    duration: const Duration(milliseconds: 500), vsync: this);
Animation<int> alpha = IntTween(begin: 0, end: 255).animate(controller);
```

注意：animate 方法返回的是一个 Animation，而不是一个 Animatable。

以下示例构建了一个控制器、一条曲线和一个 Tween，代码如下所示：

```
AnimationController controller = AnimationController(
    duration: const Duration(milliseconds: 500), vsync: this);
final Animation curve = CurvedAnimation(parent: controller, curve: Curves.easeOut);
Animation<int> alpha = IntTween(begin: 0, end: 255).animate(curve);
```

5.1.6　Curve

在 Flutter 中，通过曲线(Curve)来描述动画过程，其可以是线性的(Curves.linear)，也可以是非线性的(non-linear)。因此，整个动画过程可以是匀速的、加速的、先加速后减速的，等等。我们看一下曲线的创建方式，代码如下所示：

```
final CurvedAnimation curve = CurvedAnimation(parent: controller, curve: Curves.easeIn);
```

CurvedAnimation 和 AnimationController 都属于 Animation 类型，CurvedAnimation 可以包装 AnimationController 和 Curve 生成一个新的动画对象。我们看一个例子，代码如下所示：

```dart
import 'package:flutter/material.dart';

class CurvedAnimationPage extends StatefulWidget {
  CurvedAnimationPage({Key key, this.title}) : super(key: key);

  final String title;

  @override
  _CurvedAnimationState createState() => _CurvedAnimationState();
}

class _CurvedAnimationState extends State<CurvedAnimationPage>
    with SingleTickerProviderStateMixin {
  Animation<double> animation;
  AnimationController controller;

  @override
  void initState() {
    super.initState();
    controller = AnimationController(
        duration: const Duration(milliseconds: 2000), vsync: this);
    animation = CurvedAnimation(parent: controller, curve: Curves.bounceIn)
      ..addStatusListener((status) {
        if (status == AnimationStatus.completed) {
          controller.reverse();
        } else if (status == AnimationStatus.dismissed) {
          controller.forward();
        }
      });

    controller.forward();
  }

  dispose() {
    controller.dispose();
    super.dispose();
```

```dart
  }

  @override
  Widget build(BuildContext context) {
    return AnimatedLogo(animation: animation);
  }
}

class AnimatedLogo extends AnimatedWidget {
  // The Tweens are static because they don't change.
  static final _opacityTween = Tween<double>(begin: 0.1, end: 1.0);
  static final _sizeTween = Tween<double>(begin: 0.0, end: 300.0);

  AnimatedLogo({Key key, Animation<double> animation})
      : super(key: key, listenable: animation);

  Widget build(BuildContext context) {
    final Animation<double> animation = listenable;
    return Scaffold(
      appBar: AppBar(
        title: Text("曲线动画"),
      ),
      body: Center(
        child: Opacity(
            opacity: _opacityTween.evaluate(animation),
            child: Container(
              margin: EdgeInsets.symmetric(vertical: 10.0),
              height: _sizeTween.evaluate(animation),
              width: _sizeTween.evaluate(animation),
              child: FlutterLogo(),
            )),
      ));
  }
}
```

上述曲线运动的效果和前文 Tween 缩放的例子有点类似。我们用 CurvedAnimation 代替 Tween 方式的 Animation，并且在 CurvedAnimation 的 curve 参数里传入 Curves.bounceIn，实现了一种跳跃式的缩放效果。其曲线运动轨迹如图 5.2 所示。

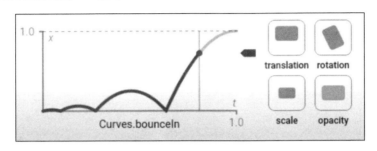

图 5.2

还有其他很多种曲线运动的效果，我们可以查看 Curves 类里的源码，比如 easeInOut，fastOutSlowIn 等。源码注释里有相关的视频效果网址，读者可以查看。

完整的代码见 chapter5/flutter_animations/lib/curved_animation.dart。

5.2 动画的封装与简化

在上一节中我们学习了动画的基本概念，了解了 Animation、AnimationController、Tween、Curve 等是如何对动画进行抽象的，也通过代码掌握了如何给动画增加监听和设置状态（addListener 和 setState）。

但是，这些是否是必需的呢？有没有简单的、经过封装的类呢？带着这些疑问，我们开始本节的学习。

5.2.1 AnimatedWidget

在上节的示例中，都是通过 addListener 和 setState 来更新 UI 的，然而有的时候可以不用那么麻烦，通过 AnimatedWidget 这个类就可以实现。它对 addListener 和 setState 进行了封装，隐藏了实现细节。

下面看一下官方为我们准备的一个 AnimatedWidget 例子，代码如下所示：

```
class AnimatedLogo extends AnimatedWidget {
  AnimatedLogo({Key key, Animation<double> animation})
      : super(key: key, listenable: animation);

  Widget build(BuildContext context) {
    final Animation<double> animation = listenable;
    return Center(
      child: Container(
        margin: EdgeInsets.symmetric(vertical: 10),
        height: animation.value,
        width: animation.value,
        child: FlutterLogo(),
      ),
    );
  }
}
```

上面经过封装后的代码精简了许多，这是因为 AnimatedLogo 可以通过当前自身 Animation 的 value 值来绘制自己。

5.2.2 AnimatedBuilder

有的时候我们会使用多个 AnimatedWidget，而如果多次实现 AnimatedWidget，则代码就显得不美观，这个时候我们就需要考虑使用 AnimatedBuilder。

AnimatedBuilder 有以下特点：

（1）不需要知道如何渲染组件，也不需要知道如何管理动画对象；

（2）继承于 AnimatedWidget，可以直接当作组件来使用，且不用显式地去添加帧的监听 addListener(...)，然后再调用 setState；

（3）只调用动画组件中的 build，在复杂布局下性能有所提高。

我们来看一下组件数示意图，如图 5.3 所示。

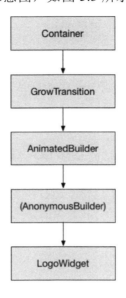

图 5.3

根据图 5.3，我们对 5.2.1 中的例子进行重构，代码如下所示：

```
import 'package:flutter/material.dart';

class AnimatedBuilderPage extends StatefulWidget {
  AnimatedBuilderPage({Key key, this.title}) : super(key: key);

  final String title;

  @override
  _AnimatedBuilderState createState() => _AnimatedBuilderState();
}

class _AnimatedBuilderState extends State<AnimatedBuilderPage>
    with SingleTickerProviderStateMixin {
  Animation<double> animation;
  AnimationController controller;

  @override
```

```dart
void initState() {
  super.initState();
  controller = AnimationController(
      duration: const Duration(milliseconds: 2000), vsync: this);
  final CurvedAnimation curve =
      CurvedAnimation(parent: controller, curve: Curves.bounceIn);
  animation = Tween(begin: 0.0, end: 300.0).animate(curve)
    ..addStatusListener((status) {
      if (status == AnimationStatus.completed) {
        controller.reverse();
      } else if (status == AnimationStatus.dismissed) {
        controller.forward();
      }
    });
  controller.forward();
}

dispose() {
  controller.dispose();
  super.dispose();
}

@override
Widget build(BuildContext context) {
  return GrowTransition(child: LogoWidget(), animation: animation);
}
}

class LogoWidget extends StatelessWidget {
  build(BuildContext context) {
    return Container(
        margin: EdgeInsets.symmetric(vertical: 10.0), child: FlutterLogo());
  }
}

class GrowTransition extends StatelessWidget {
  GrowTransition({this.child, this.animation});
```

```
    final Widget child;
    final Animation<double> animation;

    Widget build(BuildContext context) {
      return Scaffold(
        appBar: AppBar(
          title: Text("曲线动画 AnimatedBuilder 实现"),
        ),
        body: Center(
          child: AnimatedBuilder(
            animation: animation,
            builder: (BuildContext context, Widget child) {
              return Container(
                height: animation.value,
                width: animation.value,
                child: child);
            },
            child: child),
        ));
    }
}
```

读者对上述代码可能有感到困惑的地方，感觉 child 好像被指定了两次，括号外面和里面分别各有一个。其实，外面的 child 是传给 AnimatedBuilder 的，而 AnimatedBuilder 又将这个 child 作为参数传递给了里面的匿名类（AnonymousBuilder），最后 AnimatedBuilder 被插入和被渲染在组件的中间位置。

完整的代码见 chapter5/flutter_animations/lib/animatedbuilder.dart。

在 Flutter 中，通过 AnimatedBuilder 方式封装了很多动画，比如 SizeTransition、ScaleTransition、RotationTransition、FadeTransition、FractionalTranslation 等。很多时候，我们可以反复使用这些预置的过渡类，读者可以查看这些类的源码并自行探索。

5.3 Hero动画

在 Android 开发中,Shared Element Transition 可以让 Activity 或 Fragment 做出流畅的动画。同样,在 Flutter 开发中,Hero 动画也实现了类似的效果。通过 Hero 动画,我们可以在路由之间做出流畅的转场动画。

5.3.1 基本用法

Hero 的组件需要同时定义源组件和目标组件,其中源组件和目标组件被 Hero 包裹在需要动画控制的组件的外面。如果有一方不指定,在有些情况下界面就会被卡死,我们用一个例子来说明。

首先,我们在路由 1 的界面中定义出需要使用 Hero 的组件,代码如下所示:

```
class CustomLogo extends StatelessWidget {
  final double size;

  CustomLogo({this.size = 200.0});

  @override
  Widget build(BuildContext context) {
    return Container(
      color: Colors.black26,
      width: size,
      height: size,
      child: Center(
        child: FlutterLogo(
          size: 150.0,
        ),
      ),
    );
  }
}
```

这是一个自定义的组件，运行后默认效果如图 5.4 所示。

图 5.4

然后，我们在路由 2 的界面中定义出一个和前面 CustomLogo 差不多的组件，其中尺寸设定得不一样，是为了说明 Hero 产生效果的特性。

在路由 1 的界面中，我们把 CustomLogo 的外面包一层 Hero，代码如下所示：

```
child: Hero(
  tag: 'hero1',
  child: CustomLogo(
    size: 200.0,
  ),
),
```

在路由 2 的界面中，我们把 CustomLogo 的外面包一层 Hero，代码如下所示：

```
child: Hero(
  tag: 'hero1',
  child: CustomLogo(
    size: 300.0,
  ),
),
```

这样，当我们通过路由从界面 1 跳转到界面 2 时，就可以看见 Hero 产生的效果。从实际应用上来看，只需要定义 Hero 且标记其中一样就可以。

完整的代码见 chapter5/flutter_animations/lib/hero_animation.dart。

5.3.2 实现原理

整个 Hero 的运动过程分为 3 个步骤，即动画开始（t=0.0）、动画进行中、动画结束（t=1.0），5.3.1 节的例子中 Hero 的运动效果如图 5.5 所示。

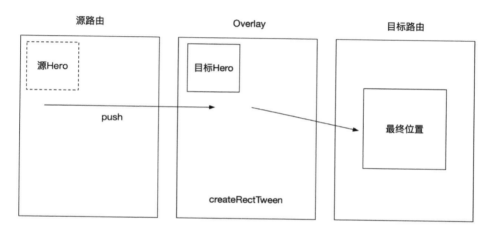

图 5.5

在图 5.5 中，我们发现在两个界面中间夹了一个 Overlay 层。在动画开始时，Flutter 会计算出 Hero 的位置并复制一份，然后绘制到 Overlay 上。复制的 Hero 和源 Hero 的大小是一致的，并且该 Hero 是在所有路由之上的。在动画实现的过程中，Flutter 会逐渐把源 Hero 移出屏幕。在动画进行中 Flutter 依靠 Tween 来实现，通过 createRectTween 属性把 Tween 传给 Hero。Hero 内部默认使用 MaterialRectArcTween 的曲线路径进行移动动画的操作。在动画结束时，Flutter 将 Overlay 中的 Hero 移除，且完成了 Hero 在目标路由上的显示，这时 Overlay 是空白的。

Hero 中所有的变换都是通过 HeroController 来实现的。HeroController

是在 MaterialApp 中通过 initState 和 didUpdateWidget 方法来完成初始化的，代码如下所示：

```
class _MaterialAppState extends State<MaterialApp> {
  HeroController _heroController;

  @override
  void initState() {
    super.initState();
    _heroController = HeroController(createRectTween: _createRectTween);
    _updateNavigator();
  }

  @override
  void didUpdateWidget(MaterialApp oldWidget) {
    super.didUpdateWidget(oldWidget);
    if (widget.navigatorKey != oldWidget.navigatorKey) {
      _heroController = HeroController(createRectTween: _createRectTween);
    }
    _updateNavigator();
  }

  RectTween _createRectTween(Rect begin, Rect end) {
    return MaterialRectArcTween(begin: begin, end: end);
  }
  // ...
}
```

在初始化 HeroController 时，Flutter 携带了一个参数，就是 _createRectTween，该参数返回的默认项就是 MaterialRectArcTween。Flutter 源码里还为我们实现了第二种 RectTween 返回值，即 MaterialRectCenterArcTween。由此可见，可以对 createRectTween 进行自定义。我们先查看 HeroController 的具体内容，代码如下所示：

```
  @override
  void didPush(Route<dynamic> route, Route<dynamic> previousRoute) {
    assert(navigator != null);
```

```
        assert(route != null);
        _maybeStartHeroTransition(previousRoute, route, HeroFlight
Direction.push, false);
    }

    @override
    void didPop(Route<dynamic> route, Route<dynamic> previousRoute) {
        assert(navigator != null);
        assert(route != null);
        _maybeStartHeroTransition(route, previousRoute,
HeroFlightDirection.pop, false);
    }
```

HeroController 其实继承的是 NavigatorObserver。在路由操作的 didPush 和 didPop 回调方法里，可以调用 _maybeStartHeroTransition，并通过 WidgetsBinding 把源路由、目标路由、HeroController 关联起来。在使用 didPush 和 didPop 回调方法时，通过调用 _startHeroTransition 方法让 Hero 动起来，只不过前者是正向的，后者是逆向的。

5.4 交错动画

Staggered Animations 即交错动画。

在现实世界中，有时候一个动画可能并不是由单一方式呈现的。渐变、位移、缩放等都是基础的动画，我们有时候需要把这些基础的动画组合起来，使其成为一个"多功能"的动画。在 Android 开发中，是通过 AnimationController 来实现这种交错动画的。在 Flutter 开发中，也沿用了名为 AnimationController 的类来实现。

交错动画举例

对于交错动画，我们可以理解为通过 AnimationController 把很多个动画对象组合在一起形成的动画，同时需要为每一段动画设置时间间隔。

对于 AnimationController 来说，控制器的值 Tween 必须属于(0.0,1.0)。根据实际情况，我们可以模拟交错动画的运动步骤，并绘制交错动画的执行顺序，如图 5.6 所示。

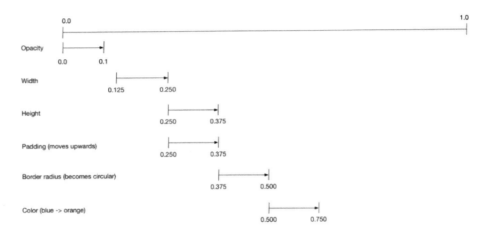

图 5.6

有了这样完整的思路之后，我们就可以做出一些自己想要实现的动画效果了。把图 5.6 的交错动画变成具体的代码，如下所示：

```
import 'package:flutter/material.dart';

import 'dart:async';
import 'package:flutter/scheduler.dart' show timeDilation;

class StaggerAnimation extends StatelessWidget {
  StaggerAnimation({Key key, this.controller}):

      opacity = Tween<double>(
        begin: 0.0,
        end: 1.0,
      ).animate(
        CurvedAnimation(
          parent: controller,
          curve: Interval(
            0.0,
```

```
        0.100,
        curve: Curves.ease,
      ),
    ),
  ),
  width = Tween<double>(
    begin: 50.0,
    end: 150.0,
  ).animate(
    CurvedAnimation(
      parent: controller,
      curve: Interval(
        0.125,
        0.250,
        curve: Curves.ease,
      ),
    ),
  ),
  height = Tween<double>(begin: 50.0, end: 150.0).animate(
    CurvedAnimation(
      parent: controller,
      curve: Interval(
        0.250,
        0.375,
        curve: Curves.ease,
      ),
    ),
  ),
  padding = EdgeInsetsTween(
    begin: const EdgeInsets.only(bottom: 16.0),
    end: const EdgeInsets.only(bottom: 75.0),
  ).animate(
    CurvedAnimation(
      parent: controller,
      curve: Interval(
        0.250,
        0.375,
        curve: Curves.ease,
```

```
        ),
      ),
    ),
    borderRadius = BorderRadiusTween(
      begin: BorderRadius.circular(4.0),
      end: BorderRadius.circular(75.0),
    ).animate(
      CurvedAnimation(
        parent: controller,
        curve: Interval(
          0.375,
          0.500,
          curve: Curves.ease,
        ),
      ),
    ),
    color = ColorTween(
      begin: Colors.indigo[100],
      end: Colors.orange[400],
    ).animate(
      CurvedAnimation(
        parent: controller,
        curve: Interval(
          0.500,
          0.750,
          curve: Curves.ease,
        ),
      ),
    ),
    super(key: key);

final Animation<double> controller;
final Animation<double> opacity;
final Animation<double> width;
final Animation<double> height;
final Animation<EdgeInsets> padding;
final Animation<BorderRadius> borderRadius;
final Animation<Color> color;
```

```dart
    // This function is called each time the controller "ticks" a
new frame.
    // When it runs, all of the animation's values will have been
    // updated to reflect the controller's current value.
    Widget _buildAnimation(BuildContext context, Widget child) {
      return Container(
        padding: padding.value,
        alignment: Alignment.bottomCenter,
        child: Opacity(
          opacity: opacity.value,
          child: Container(
            width: width.value,
            height: height.value,
            decoration: BoxDecoration(
              color: color.value,
              border: Border.all(
                color: Colors.indigo[300],
                width: 3.0,
              ),
              borderRadius: borderRadius.value,
            ),
          ),
        ),
      );
    }

    @override
    Widget build(BuildContext context) {
      return AnimatedBuilder(
        builder: _buildAnimation,
        animation: controller,
      );
    }
  }

  class StaggerDemo extends StatefulWidget {
```

```dart
  @override
  _StaggerDemoState createState() => _StaggerDemoState();
}

class _StaggerDemoState extends State<StaggerDemo>
    with TickerProviderStateMixin {
  AnimationController _controller;

  @override
  void initState() {
    super.initState();

    _controller = AnimationController(
        duration: const Duration(milliseconds: 2000), vsync: this);
  }

  @override
  void dispose() {
    _controller.dispose();
    super.dispose();
  }

  Future<void> _playAnimation() async {
    try {
      await _controller.forward().orCancel;
      await _controller.reverse().orCancel;
    } on TickerCanceled {
      // the animation got canceled, probably because we were disposed
    }
  }

  @override
  Widget build(BuildContext context) {
    timeDilation = 10.0; // 1.0 is normal animation speed.
    return Scaffold(
      appBar: AppBar(
        title: const Text('Staggered Animation'),
      ),
```

```
      body: GestureDetector(
        behavior: HitTestBehavior.opaque,
        onTap: () {
          _playAnimation();
        },
        child: Center(
          child: Container(
            width: 300.0,
            height: 300.0,
            decoration: BoxDecoration(
              color: Colors.black.withOpacity(0.1),
              border: Border.all(
                color: Colors.black.withOpacity(0.5),
              ),
            ),
            child: StaggerAnimation(controller: _controller.view),
          ),
        ),
      ),
    );
  }
}
```

完整的代码见 chapter5/flutter_animations/lib/stagger_demo.dart。

5.5 动画示例

在本节中，我们结合实际项目中用到的例子讲解一下动画，并通过例子讲解和回顾一下本章所学知识。

5.5.1 自定义加载动画

我们将本章所学的内容通过一个完整的自定义 Loading 加载框来呈现。在实际项目实践当中，加载框肯定是存在的。通过对本例的学习，希望读者能举一反三，绘制出自己想要的加载框效果。本例完成之后的效果如图 5.7 所示。

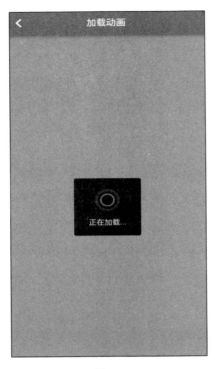

图 5.7

下面来说一下实现思路：

（1）自定义带动画效果的组件，即图中会动的、重复的、淡入淡出的波纹效果；

（2）通过 Stack 布局实现一个 Dialog 对话框效果，居中显示，并且在耗时任务完成时可以关闭；

（3）把前面实现的 Dialog 效果、自定义动画组件和加载的文字提示结合起来，形成一个加载组件。

下面，我们分小节讲解一下具体的实现过程。

5.5.2 实现动画效果

从图 5.7 的效果演示可以看到，运行起来会有两个圆圈，从小到大向外扩散，扩散到最后就是淡出效果，如此反复循环。我们构造出两个 Animation 的动画对象，通过 Tween 传入 CurvedAnimation，代码如下所示：

```
_animation1 = Tween(begin: 0.0, end: 1.0).animate(
  CurvedAnimation(
    parent: _controller,
    curve: const Interval(0.0, 0.75, curve: Curves.linear),
  ),
)..addListener(() => setState(() => <String, void>{}));

_animation2 = Tween(begin: 0.0, end: 1.0).animate(
  CurvedAnimation(
    parent: _controller,
    curve: const Interval(0.25, 1.0, curve: Curves.linear),
  ),
)..addListener(() => setState(() => <String, void>{}));
```

然后，我们创建 Stack 布局组件，把两个圆圈重叠在一起，代码如下所示：

```
children: <Widget>[
  Opacity(
    opacity: 1.0 - _animation1.value,
    child: Transform.scale(
      scale: _animation1.value,
      child: _itemBuilder(0),
    ),
  ),
  Opacity(
    opacity: 1.0 - _animation2.value,
    child: Transform.scale(
      scale: _animation2.value,
      child: _itemBuilder(1),
```

```
      ),
    ),
  ],
```

相信大家看到这里已经明白了该动画的原理,后续代码如下所示:

```
import 'package:flutter/material.dart';

class SpinKitRipple extends StatefulWidget {
  const SpinKitRipple({
    Key key,
    this.color,
    this.size = 50.0,
    this.borderWidth = 6.0,
    this.itemBuilder,
    this.duration = const Duration(milliseconds: 1800),
  }) : assert(color != null),
        assert(size != null),
        assert(borderWidth != null),
        super(key: key);

  final Color color;
  final double size;
  final double borderWidth;
  final IndexedWidgetBuilder itemBuilder;
  final Duration duration;

  @override
  _SpinKitRippleState createState() => _SpinKitRippleState();
}

class _SpinKitRippleState extends State<SpinKitRipple>
    with TickerProviderStateMixin {
  AnimationController _controller;
  Animation<double> _animation1, _animation2;

  @override
  void initState() {
    super.initState();
```

```
    _controller = AnimationController(vsync: this, duration:
widget.duration)
      ..repeat();

    _animation1 = Tween(begin: 0.0, end: 1.0).animate(
      CurvedAnimation(
        parent: _controller,
        curve: const Interval(0.0, 0.75, curve: Curves.linear),
      ),
    )..addListener(() => setState(() => <String, void>{}));

    _animation2 = Tween(begin: 0.0, end: 1.0).animate(
      CurvedAnimation(
        parent: _controller,
        curve: const Interval(0.25, 1.0, curve: Curves.linear),
      ),
    )..addListener(() => setState(() => <String, void>{}));
  }

  @override
  void dispose() {
    _controller.dispose();
    super.dispose();
  }

  @override
  Widget build(BuildContext context) {
    return Container(
      child: Stack(
        children: <Widget>[
          Opacity(
            opacity: 1.0 - _animation1.value,
            child: Transform.scale(
              scale: _animation1.value,
              child: _itemBuilder(0),
            ),
          ),
          Opacity(
```

```
            opacity: 1.0 - _animation2.value,
            child: Transform.scale(
              scale: _animation2.value,
              child: _itemBuilder(1),
            ),
          ),
        ],
      ),
    );
  }

  Widget _itemBuilder(int index) {
    return SizedBox.fromSize(
      size: Size.square(widget.size),
      child: widget.itemBuilder != null
          ? widget.itemBuilder(context, index)
          : DecoratedBox(
              decoration: BoxDecoration(
                shape: BoxShape.circle,
                border:
                    Border.all(color: widget.color, width: widget.borderWidth),
              ),
            ),
    );
  }
}
```

运行整个代码，看一下最终效果，如图 5.8 所示。

图 5.8

到这里一个简单的动画效果已经完成了。

5.5.3　Dialog 加载框

构建对话框的思路其实和做 Android 或 iOS 相关功能的思路是一样的。我们可以对 Dialog 封装加载文字内容、透明度及颜色，为前面自定义的动画组件嵌入插槽。完整代码如下所示：

```
import 'package:flutter/material.dart';
import 'package:flutter_animations/spin_kit_ripple.dart';

class ProgressDialog extends StatelessWidget {
  //子布局
  final Widget child;

  //加载中是否显示
  final bool isLoading;

  //进度提醒内容
  final String message;

  //加载中动画
  final Widget progress;

  // 加载框背景透明度
  final double alpha;

  // 字体颜色
  final Color textColor;

  ProgressDialog(
      {Key key,
      @required this.isLoading,
      this.message,
      this.progress = const SpinKitRipple(color: Colors.white, size: 60.0),
      this.alpha = 0.5,
```

```dart
      this.textColor = Colors.white,
    @required this.child})
    : assert(child != null),
      assert(isLoading != null),
      super(key: key);

  @override
  Widget build(BuildContext context) {
    List<Widget> widgetList = [];
    widgetList.add(child);
    if (isLoading) {
      Widget layoutProgress;
      if (message == null) {
        layoutProgress = Center(
          child: progress,
        );
      } else {
        layoutProgress = Center(
          child: Container(
            padding: const EdgeInsets.all(20.0),
            decoration: BoxDecoration(
              color: Colors.black87,
              borderRadius: BorderRadius.circular(4.0),
            ),
            child: Column(
              mainAxisAlignment: MainAxisAlignment.center,
              crossAxisAlignment: CrossAxisAlignment.center,
              mainAxisSize: MainAxisSize.min,
              children: <Widget>[
                progress,
                Container(
                  padding: const EdgeInsets.fromLTRB(10.0, 10.0, 10.0, 0),
                  child: Text(
                    message,
                    style: TextStyle(color: textColor, fontSize: 16.0),
                  ),
                )
              ],
```

```
          ),
        ),
      );
    }
    widgetList.add(Opacity(
      opacity: alpha,
      child: new ModalBarrier(color: Colors.black87),
    ));
    widgetList.add(layoutProgress);
  }
  return Stack(
    children: widgetList,
  );
 }
}
```

在上述代码中,我们定义了一个 Widget,然后初始化前文所讲的自定义动画。代码如下所示:

```
this.progress = const SpinKitRipple(color: Colors.white, size: 60.0),
```

我们也能任意修改文字的内容,并且 Dialog 的动画和文字可以是上下结构的,也可以是左右结构的,我们可以根据实际情况进行布局。

5.5.4 测试加载框效果

最后,我们加入以下测试代码调用自定义加载框,代码如下所示:

```
import 'package:flutter/material.dart';
import 'package:flutter_animations/progress_dialog.dart';

class LoadingPage extends StatefulWidget {
  LoadingPage({Key key, this.title}) : super(key: key);

  final String title;
```

```dart
  @override
  _LoadingState createState() => _LoadingState();
}

class _LoadingState extends State<LoadingPage> {
  bool _loading = false;

  @override
  Widget build(BuildContext context) {
    return Scaffold(
      appBar: AppBar(
        title: Text("加载动画"),
      ),
      body: ProgressDialog(
        isLoading: _loading,
        message: '正在加载...',
        alpha: 0.35,
        child: Center(
          child: RaisedButton(
            onPressed: () => _onRefresh(),
            child: Text('显示加载动画'),
          ),
        ),
      ),
    );
  }

  Future<Null> _onRefresh() async {
    setState(() {
      _loading = !_loading;
    });
    // 模拟耗时操作
    await Future.delayed(Duration(seconds: 5), () {
      setState(() {
```

```
        _loading = !_loading;
      });
    });
  }
}
```

这样，整个自定义动画的加载框就完成了。如果对于动画效果不满意，想换成别的也行，只需要在实现别的动画效果后传入加载框即可。

完整的代码见 chapter5/flutter_animations/lib/loading.dart。

本章小结

本章讲解了动画的相关知识和原理，也尝试了不同种类动画的实现方法，希望读者勤加练习，并在工作实践中做出想要的动画效果。本章还有一个未提及的内容，就是路由自定义动画的实现，笔者会在路由章节中进行详细讲解。在 Flutter 1.2 版本更新之后，官方文档又对动画这一章进行了更新，加入了更多的视频讲解。

第 6 章

使用网络技术与异步编程

本章会先给大家介绍一下网络协议相关的一些基础知识（有相关基础的读者可以跳过），然后会给大家讲解 Flutter 怎样通过网络来获取数据。说到网络操作就想起异步编程，有过 Android 开发经验的同学一定知道，网络操作是不能放在 UI 主线程里的，它必须通过异步的方式来实现，本章将会介绍 Flutter 的异步编程。网络操作和异步编程是 Flutter 里需要重点掌握的知识内容。

6.1 网络协议简介

网络协议简单来说就是为计算机网络进行数据交换而建立的规则、标准或约定的集合。

常用的网络协议有 TCP/IP 协议、HTTP 协议、FTP 协议、Telnet 协议、SMTP 协议、NFS 协议、UDP 协议等。

一名优秀的移动端开发工程师，一定对网络协议有比较深入的了解，网络协议被广泛运用在互联网的各个领域，包括前端、后端等。本节将介绍一下网络协议的基本情况，如果你对网络协议比较了解的话，可以跳过这一节。

6.1.1　HTTP 协议简介

HTTP 协议是互联网的基础协议，是基于 TCP/IP 协议的应用层协议。它通过客户端发送请求给服务端，然后服务端再响应给浏览器，如图 6.1 所示。我们平时在访问各种网页时就会用到 HTTP 协议。

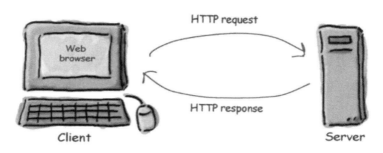

图 6.1

HTTP 协议从早期的 HTTP 0.9 版本（只能返回 HTML）发展至今，已经产生了很大的变化，最新的版本 HTTP 2.0 更是让它成了一个技术热点。

目前我们看到的各种在线资源，很多都是基于 HTTP 协议传输的。服务端在返回信息给客户端时会告诉客户端用的是什么格式。另外，我们在查看 Response 里的信息时会发现有一个名为 Content-Type 的字段，这个字段包含了 MIME type，MIME type 常见的数据类型如下：

- text/plain

- text/html

- text/css

- image/jpeg

- image/png

- image/svg+xml

- audio/mp4

- video/mp4

- application/javascript

- application/pdf

- application/zip

- application/atom+xml

读者可以尝试下载 Chrome 浏览器，随便输入一个网址，然后在调试面板里打开 Network 选项，再打开 Headers 选项，在 Response Headers 里面可以看见类似"Content-Type: application/javascript"这样的信息，说明这种情况下返回的是 JavaScript。如果是图片则会显示"Content-Type: image/gif"，Content-Type 也支持自定义。

其他一些常用字段的相关概念我们也需要了解，因为这对于解决一些技术问题有很大帮助，具体如下所示。

（1）Content-Encoding：该字段说明的是数据的压缩方式，常见的有 gzip 压缩。

（2）Connection：在 HTTP 1.1 协议里，默认开启了 Connection 的 keep-alive，也是 HTTP 1.1 协议带来的最大变化。它指的是持久连接，即 TCP 连接在默认情况下不关闭，能被多个请求复用。

（3）Content-Length：表示数据长度，在 HTTP 1.1 里不是必需的，支持分块传输，即产生一块数据就发送一块，如设置了 Transfer-Encoding，则表示服务的回应由数量未定的数据块组成。

（4）Status：状态码，常见的有 200、304、400、404、500 等。

6.1.2　HTTP 2.0 能给我们带来什么

在前面小节中我们提到了 HTTP 1.1 里面的 Connection: keep-alive，TCP 支持复用，但是同一个 TCP 里的数据通信只支持依次执行，这是 HTTP 1.1 明显的一个缺陷，直到 HTTP 2.0 才得以修复。

HTTP 2.0 给我们带来的最大的改善就是实现了 HTTP 的多路复用，我们先来看一张图，如图 6.2 所示。

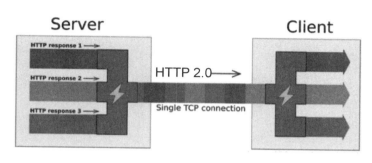

图 6.2

图 6.2 中 Client 多路复用了 TCP 连接。在一个 TCP 连接的情况下，客户端和浏览器可以同时发送多个请求和响应，实现了双向且实时的通信。

HTTP 2.0 的数据包不是按顺序发送的。一个连接中连续的数据包，可能归属于不同的回应。因此，在 HTTP 2.0 里有了数据流（Stream）的概念。每一个数据流都是唯一的，用 ID 来标识，它指向不同的回应，起到区分的作用。客户端发送的数据流的 ID 为奇数，服务器发送的数据流的 ID 为偶数。并且在数据流未完成发送时，可以取消。HTTP2.0 可以在 TCP 连接还在打开的情况下，取消某一次请求，以便 TCP 可以被其他请求使用。数据流还可以设置优先级，优先级高的服务器就会优先响应。

HTTP 2.0 其他的显著特性如下所示。

1. 二进制协议

HTTP 2.0 采用彻底的二进制协议，头信息和数据体都是二进制的（即头信息帧和数据帧），它带来的好处是，解析数据变得更加简单、方便。

2. 头信息压缩机制（Header Compression）

头信息使用 gzip 或 compress 压缩后再发送，客户端和服务器之间会生成一个索引号，用于维护头信息表。头信息表里存放着请求头（Request Header）的信息，这样避免了 HTTP 1.1 里每次都携带相同的头信息的情况，从而只需要发送索引号，并提升了速度。

3. 服务器推送

一般而言，客户端在请求服务器之后，会解析 HTML 资源，如果发现有静态资源会再发出静态资源的请求。在 HTTP 2.0 里，可以预判客户端需要请求服务端的资源，并主动把这些资源随着网页一起发送给客户端，下面这张图展示了服务端推送的原理，如图 6.3 所示。

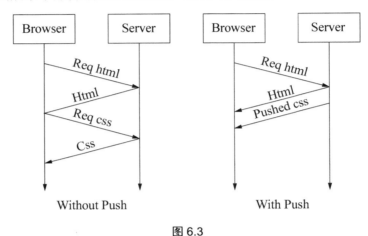

图 6.3

6.1.3 HTTPS

简单来说，HTTPS 就是安全的 HTTP，即 HTTP+SSL。HTTPS 集合了加密传输、身份认证，因此更加安全。

HTTPS 与 HTTP 的区别，如下所示。

（1）HTTPS 的服务器需要申请 CA 证书。

（2）HTTP 传输的信息是明文的，HTTPS 传输的信息是密文的。

（3）HTTP 的端口是 80，HTTPS 的端口是 443。

前面说了，HTTPS 是加密传输的，且经历了握手（Handshake）的过程。

HTTPS 握手的大致过程如下所示。

（1）客户端请求建立 SSL 链接，并向服务端发送一个随机数（Client random）和客户端支持的加密方法（比如 RSA 公钥加密），此时是明文传输的。

（2）服务端回复一种客户端支持的加密方法、一个随机数（Server random）、授信的服务器证书和非对称加密的公钥。

（3）客户端收到服务端的应答后则使用服务端的公钥，加上新的随机数（Premaster secret）以及从服务端下发到客户端的公钥及加密方法进行加密，再发送给服务器。

（4）服务端收到客户端的回复后，利用已知的加密/解密方式进行解密。同时，在一定的算法下利用 Client random、Server random 和 Premaster secret，生成 HTTP 链接数据传输的对称加密密钥。

6.2 网络编程

Android 常用的网络库有以下几种：HttpClient、HttpUrlConnection、Volley、Okhttp、Retrofit。对于 Android 来说，使用最多的网络库应该是 Okhttp 和 Retrofit 的结合。那么，我们看一下 Flutter 网络库的发展情况。

Flutter 获取数据最常用的方式有两种，即官方自带的 HttpClient 和 http 库。下面我们分别来介绍一下各自的用法。

6.2.1 HttpClient

作为 Dart 语言的一部分，HttpClient 是 Dart 原生的网络请求/获取数据的方式。下面，我们看一下 Dart 官方为我们提供的 get 请求的例子，代码结构如下所示：

```
HttpClient client = new HttpClient();
client.getUrl(Uri.parse("http://www.example.com/"))
    .then((HttpClientRequest request) {
```

```
    // Optionally set up headers...
    // Optionally write to the request object...
    // Then call close.
    ...
    return request.close();
  })
  .then((HttpClientResponse response) {
    // Process the response.
    ...
  });
```

结合以上代码和前文所学我们可以知道，在初始化 HttpClient 之后（HttpClient 是一个抽象类），程序调用了 client.getUrl 的方法，该方法的返回值是 Future，因此可以通过.then 的方式进行链式调用。我们改造一下上述代码，用 async/await 作为"语法糖"来实现一下，并写一个例子出来，部分代码如下所示：

```
_loadData() async {
  try {
    HttpClient httpClient = new HttpClient();
    HttpClientRequest request =
        await httpClient.getUrl(Uri.parse ("https://www.phei.com.cn"));
    HttpClientResponse response = await request.close();
    _responseText = await response.transform(Utf8Decoder()).join();
    print(_responseText);
    httpClient.close();
  } catch (_) {
    print('请求异常: ' + _.toString());
  }
}
```

在上述代码中，我们请求了一个网址，在通过一个按钮触发_loadData 方法之后，会在控制台打印出请求发出以后所响应回来的内容，如图 6.4 所示。

```
<meta http-equiv="Content-Type" content="text/html; charset=utf-8">
<meta http-equiv = "X-UA-Compatible" content = "IE=edge,chrome=1" />
<title>电子工业出版社有限公司</title>
<meta name="keywords" content="" />
<meta name="description" content="" />
<link rel="shortcut icon" href="#" />
<link href="/templates/stylesheets/global.css" rel="stylesheet" />
<link href="/templates/stylesheets/web_base.css" rel="stylesheet" />
<script src="/templates/javascript/jquery-3.2.1.min.js" type="text/javascript"></script>
<script src="/templates/javascript/ajax.js" type="text/javascript"></script>
<script src="/templates/javascript/page_script.js" type="text/javascript"></script>
<script src="/templates/javascript/angular/angular.min.js" type="text/javascript"></script>
```

图 6.4

在 HttpClient 中，通过常规的 get 方法来请求网络就是如此了。需要注意的是，HttpClient 在使用之后一定要调用 close 方法来关闭。在 close 方法里有一个 force 的可选参数，如果设定为 false，则会保持当前可用网络的连接且在等待完成后才关闭；如果为 true，则会立即关闭并释放资源。

在 HttpClient 中，也能为 HttpClientRequest 设置请求头。笔者在做 Android 开发时，有的时候会在调试网络时设置断点。通过断点可以看到请求网络的 Request Header 参数具体是什么。这个参数是可以自定义的，它通常与服务端沟通/约定请求头的内容。HttpClient 默认的请求头是 Accept-Encoding: gzip，我们也可以自己设定请求头，代码如下所示：

```
request.header.add("user-agent", "请求头的内容");
```

在本例中，我们只介绍了 get 方式获取的请求。但在实际项目中，我们肯定存在与服务端的交互情况。在交互时，需要传入特定参数且不被发现参数的具体值，这个时候就可以考虑用 post 方式来获取数据。我们可以把 httpClient.getUrl 改成 httpClient.postUrl，然后根据需要传入与服务端所约定的一些参数（post 请求发送的）。举一个例子，代码如下所示：

```
_loadDataPostMethod() async {
  try {
    HttpClient httpClient = new HttpClient();
    HttpClientRequest request =
        await httpClient.postUrl(Uri.parse
("https://post.example.com"));
    request.headers.set('content-type', 'application/json');
    Map jsonMap = {'userid': '10000'};
    request.add(utf8.encode(json.encode(jsonMap)));
```

```
            HttpClientResponse response = await request.close();
            _responseText = await response.transform(Utf8Decoder()).join();
            print(_responseText);
            httpClient.close();
        } catch (_) {
            print('请求异常：' + _.toString());
        }
    }
```

完整的代码见 chapter6/flutter_network/lib/httpclient_test.dart。

6.2.2　http 库

介绍完了 HttpClient 之后，我们来看一下 Flutter 官方为我们推荐的网络请求库 http。

http 库包含了一些"高阶函数"，可以让我们更方便地访问网络，获取资源。http 库同时支持手机端和 PC 端。本节我们主要学习如何用 http 库发起 get/post 请求，如何把发出去的请求的响应结果转换成 Dart 对象，以及如何将请求结果展示在页面上。

下面，我们看一下 http 库添加依赖关系和简单的发送请求的例子。

首先，在 pubspec.yaml 里添加 http 的库，代码如下所示：

```
dependencies:
  http: ^0.12.0+1
```

在添加完库之后，使用 http 发送网络请求，我们看一下发送请求之后返回的内容，代码如下所示：

```
Future<http.Response> fetchGet() {
  return http.get('https://example.com');
}
```

我们可以改造前面 HttpClient 的例子，用 http 来实现，代码如下所示：

```
_loadData() async {
  var client = http.Client();
  var uri = Uri.parse("https://www.phei.com.cn");
  http.Response response = await client.get(uri);
  print(utf8.decode(response.bodyBytes));
  client.close();
}
```

记得要用 utf8.decode 方法把网页获取的内容进行转码,否则默认获取到的内容就是乱码形式的。

在上面的代码中,我们用到了 http 的 get 请求方式,但如果是 post 请求,则可以通过 http.post 的方法发送请求。我们参照并改写上一节中 HttpClient 的 post 请求发送方式,改写后代码如下所示:

```
_loadDataPostMethod() async {
  var client = http.Client();
  Map<String, String> headerMap = {'userid': '10000'};
  http.Response response = await client
      .post("https://www.phei.com.cn", headers: headerMap, body: {});
  print(utf8.decode(response.bodyBytes));
  client.close();
}
```

我们看一下 client.post 的源码,如下所示:

```
Future<Response> post(url,
    {Map<String, String> headers, body, Encoding encoding});
```

这个方法可以传入 3 个参数:headers、body 和 encoding,我们可以根据项目的实际需要以及和服务端的约定来进行传值调用。

完整的代码见 chapter6/flutter_network/lib/http_test.dart。

6.3　JSON解析

前文我们讲解了 HttpClient 和 http 库的基本用法,我们请求一段网页,并且返回了网页的 HTML 内容，但这样是不够的。因为我们做的是移动端开发，通常服务的请求和响应的格式是 JSON 格式，是通过服务端返回的。

JSON（JavaScript Object Notation，JS 对象简谱）是一种轻量级的数据交换格式。那么本节我们来介绍一下怎样把服务端返回的结果进行 JSON 解析。

6.3.1　JSON 转成 Dart 对象

假设，我们现在开发的是一款电影类的 App，通过访问相关接口之后，服务端返回这样一条简单的 JSON 数据，如下所示：

```
{"title": "我不是药神"}
```

在获得数据之后，相信有过 Android 或前端开发经验的读者，肯定知道要把这段数据还原成一个对象，并且在界面上展示出来。在 Flutter 里，可以把这个 JSON 对象转化成 Dart 对象。这里暂时取一个名字，就叫 Movie，可以这样写，代码如下所示：

```dart
class Movie {
  final String title;

  Movie({this.title});

  factory Movie.fromJson(Map<String, dynamic> json) {
    return Movie(
      title: json['title']
    );
  }
}
```

在 dart:convert 里有一个 JSON 常量，它是负责处理服务端返回的 JSON 数据的。

在请求响应回来时，通过 json.decode(response.body) 方法调用可以把 JSON 结果转化成 Map 类型或 List 类型。如果是一个 JSON 对象，返回值将是一个 Map；如果是 JSON 数组，则会返回 List。返回一个 dynamic 类型的原因在于 Dart 不知道传进去的 JSON 是什么数据类型。我们可以通过 json['map-key']的形式来获取它。

6.3.2 一个完整的例子

在 JSON 解析之后，我们是需要把结果展现给用户看的。最终展示的结果可能是一串文字、一张图片、一个列表。我们看一个例子，代码如下所示：

```
import 'dart:async';
import 'dart:convert';

import 'package:flutter/material.dart';
import 'package:http/http.dart' as http;

Future<Movie> fetchPost() async {
  final response =
      await http.get('http://172.20.10.3:3000/movies/detail/1');
  if (response.statusCode == 200) {
    return Movie.fromJson(json.decode(response.body));
  } else {
    throw Exception('Failed to load post');
  }
}

class Movie {
  final String title;

  Movie({this.title});

  factory Movie.fromJson(Map<String, dynamic> json) {
    return Movie(
      title: json['title']
    );
  }
```

```
  }

  class HomePage extends StatefulWidget {
    @override
    _HomePageState createState() => new _HomePageState(movie:
fetchPost());
  }

  class _HomePageState extends State<HomePage> {
    final Future<Movie> movie;

    _HomePageState({Key key, this.movie});

    @override
    Widget build(BuildContext context) {
      return new Scaffold(
        body: Center(
          child: FutureBuilder<Movie>(
            future: movie,
            builder: (context, snapshot) {
              if (snapshot.hasData) {
                return Text(snapshot.data.title);
              } else if (snapshot.hasError) {
                return Text("${snapshot.error}");
              }
              // By default, show a loading spinner
              return CircularProgressIndicator();
            },
          )
        )
      );
    }
  }
```

这是一个通过 http 库请求服务端接口，然后在服务端响应正常的情况下（response.statusCode 为 200），把 JSON 还原成一个实体类，并通过 FutureBuilder 组件把结果输出到屏幕上的例子。

完整的代码见 chapter6/flutter_network/lib/http_demo.dart。

6.3.3 根据 JSON 用工具生成实体类

在上述例子中,我们已经知道了怎样解析 JSON,然后还原成一个实体类。这种方式是手动写的,在一些复杂的业务场景中,实体类的属性可能有许多,一个一个手写太耗费时间。那么,我们能否用一些省时省力的方式,通过一些工具去生成呢?答案当然是肯定的。

官方在 pubspec.yaml 中引入了以下依赖库:

```
dependencies:
  # 加入 dependencies 依赖
  json_annotation: ^2.0.0

dev_dependencies:
  # 加入 dev_dependencies 依赖
  build_runner: ^1.0.0
  json_serializable: ^2.0.0
```

引入 annotation 需要的 3 个包之后,我们可以通过代码生成器去生成代码。我们可以创建一个类来试一下,代码如下所示:

```
import 'package:json_annotation/json_annotation.dart';

part 'user.g.dart';

@JsonSerializable()

class User {
  User(this.name, this.email);

  String name;
  String email;

  factory User.fromJson(Map<String, dynamic> json) =>
_$UserFromJson(json);
```

```
    Map<String, dynamic> toJson() => _$UserToJson(this);
}
```

为了便于记忆，我们可以把上面创建的类理解为"模板"。当发现 fromJson 和 toJson 前面加了"_$"这样的占位符时，代码是会报错的。不要着急，我们来生成一下 JSON 的 fromJson 和 toJson 方法，在工程主目录下，我们输入以下命令：

```
flutter packages pub run build_runner build
```

输入完以上命令之后，如果没报错，就会生成一个名为 user.g.dart 的文件。

这种方式是一次性生成的。有的时候，我们会对实体类进行修改，一次一次地生成显然有些烦琐，于是，我们可以通过监听的模式来实现每一次的生成，命令如下所示：

```
flutter packages pub run build_runner watch
```

在实际环境中，接口返回的数据的嵌套关系肯定比较复杂。通过上述方式，我们可以减少一些烦琐的手动编写代码过程，并且可以防止手动编写带来的错误，提高开发效率。

有的时候，我们实体类里的属性和服务端返回的 JSON 字段是不一致的，这个时候也能通过 JsonKey 注解的方式指向正确的接口字段，方式如下所示：

```
@JsonKey(name: 'user_name')
final String userName;
```

6.4 dio 库

dio 库是 Flutter 中文网提供的一个强大的 http 请求库，在 GitHub 上它的 star 数量已经超过了 3000 次。看了文档目录之后，笔者认为 dio 库和 Android 的 OkHttp 库有一些相似。它支持文件的上传/下载、Cookie 管理、FormData、请求/取消、拦截器等操作。下面，我们对常用的几个操作进行一些说明。

6.4.1 基本用法

基本用法我们就说一下 get 和 post 用法。比如获取一个网址并打印结果，可以通过下面的方式实现，代码如下所示：

```
_loadData() async {
  try {
    Response response = await Dio().get("https://www.phei.com.cn");
    print(response);
  } catch (e) {
    print(e);
  }
}
```

我们再看一下如何使用 post，代码如下所示：

```
_loadDataPostMethod() async {
  try {
    Response response = await Dio().post("https://www.phei.com.cn", data: {});
    print(response);
  } catch (e) {
    print(e);
  }
}
```

完整的代码见 chapter6/flutter_network/lib/dio_test.dart。

6.4.2 dio 单例

一个 dio 实例可以发起多个网络请求。很多时候，在 App 里只存在一个 http 数据源，且有一些公共的配置项，比如公共的头文件、公共的请求地址、超时设置等。在这些情况下，我们可以考虑采用 dio 的单例模式，这种方式也是官方推荐的，更便于统一管理。创建 dio 单例的方式如下所示：

```
var dio = new Dio(new BaseOptions(
  baseUrl: "http://www.dtworkroom.com/doris/1/2.0.0/",
  connectTimeout: 5000,
  receiveTimeout: 100000,
  // 5s
  headers: {
    HttpHeaders.userAgentHeader: "dio",
    "api": "1.0.0",
  },
  contentType: ContentType.json,
  // Transform the response data to a String encoded with UTF8.
  // The default value is [ResponseType.JSON].
  responseType: ResponseType.plain,
));
```

从上面的实例中,我们可以看到,在初始化 dio 实例时,我们传入了一个 BaseOptions 的配置项,里面可以设定一些基本且共用的信息,比如 baseUrl、connectTimeout、receiveTimeout、headers 等信息。这样更便于我们对 App 的网络请求进行统一的管理。

6.4.3 dio 拦截器

dio 库可以像 Android 里的 OkHttp 库一样,根据实际的需要,在请求之前或响应之后(还没有被 then 或 catchError 处理)做一些统一的预处理操作。我们看一下官方给的例子,代码如下所示:

```
dio.interceptors.add(InterceptorsWrapper(
  onRequest:(RequestOptions options){
    // 在请求被发送之前做一些事情
    return options; //continue
  },
  onResponse:(Response response) {
    // 在返回响应数据之前做一些预处理
    return response; // continue
  },
  onError: (DioError e) {
```

```
      // 当请求失败时做一些预处理
      return e;//continue
    }
));
```

可以看到，在拦截器做 Request 处理时，我们可以直接返回 options，这个时候会继续处理 then 方法里的逻辑。如果你想完成请求并返回一些自定义的数据，可以返回一个 Response 对象或返回 dio.resolve(data)，这样请求将会被终止，上层 then 会被调用，then 中返回的数据将是你的自定义数据 data。如果你想终止请求并触发一个错误，你可以返回一个 DioError 对象，或返回 dio.reject(errMsg)，这样请求将被中止并触发异常，上层 catchError 会被调用。比如，可以这样写，代码如下所示：

```
dio.interceptors.add(InterceptorsWrapper(
  onRequest:(RequestOptions options){
    return dio.resolve("fake data")
  },
));
Response response = await dio.get("/test");
print(response.data);   //"fake data"
```

在一些业务场景中，我们可能需要锁定/解锁拦截器，一旦请求/响应的拦截器被锁定，接下来的请求/响应，将会在进入拦截器之前进行排队等待，直到解锁成功，这些排队的请求/响应才会继续被执行（进入拦截器）。在一些需要串行化请求/响应的场景中非常实用，我们可以使用 lock/unlock 来实现，代码如下所示：

```
dio.interceptors.add(InterceptorsWrapper(
    onRequest: (Options options) {
        print('send request: path:${options.path}, baseURL:${options.baseUrl}');
        if (csrfToken == null) {
            print("no token, request token firstly...");
            //lock the dio.
            dio.lock();
            return tokenDio.get("/token").then((d) {
                options.headers["csrfToken"] = csrfToken = d.data
```

```
['data']['token'];
                print("request token succeed, value: " + d.data
['data']['token']);
                print(
                    'continue to perform request: path:$
{options.path}, baseURL:${options.path}');
                return options;
            }).whenComplete(() => dio.unlock()); // unlock the dio
        } else {
            options.headers["csrfToken"] = csrfToken;
            return options;
        }
    }
));
```

在上述例子中，我们创建一个实例用于请求 token，在发送请求时，我们会被请求的拦截器拦截。如果我们首次访问接口，在没有携带 token 的情况下，代码里会创建另一个 dio 实例并通过异步任务去获取 token，获取完成之后，dio 会把它设置到 headers 里面，并把 token 作为请求参数发送，然后调用 unlock 方法进行解锁。你也可以调用拦截器的 clear() 方法来清空等待队列。

6.4.4　dio 拦截器链

其实，拦截器不止一个。在 Android 的 OkHttp 框架里，就有"拦截器链"的概念，在 dio 框架里也一样。查看一下代码，笔者发现拦截器都是放在 ListMixin 里面的，最终需要执行_executeInterceptors 来处理每一个拦截器，代码如下所示：

```
    Future _executeInterceptors<T>(T ob, f(Interceptor inter, T
ob)) async {
        for (var inter in interceptors) {
            var res = await _assureFuture(f(inter, ob));
            if (res != null) {
                if (res is T) {
                    ob = res;
```

```
            continue;
        }
        if (res is Response || res is DioError) return res;
        return res;
    }
}
return ob;
}
```

通过学习代码，就比较容易理解它的工作原理了。我们可以根据实际的业务场景来添加多个拦截器。比如，我们可以添加日志拦截器，代码如下所示：

```
dio.interceptors.add(LogInterceptor(responseBody: false));
//开启请求日志
```

再比如，如果我们需要添加 Cookie 进行管理，也能放到拦截器里面，可以这样写，代码如下所示：

```
dio.interceptors.add(CookieManager(CookieJar()))
```

6.4.5 dio 适配器

Flutter 还能抽象出适配器来方便切换和定制底层网络库。比如，在 Flutter 中我们可以通过自定义 HttpClientAdapter 将 http 请求转发到 Native 中，然后再由 Native 统一发起请求。再比如，将来某一天 OkHttp 也提供了 Dart 版本，这个时候可以通过适配器无缝切换到 OkHttp 库，而不用修改之前的代码。dio 使用 DefaultHttpClientAdapter 作为其默认的 HttpClientAdapter，DefaultHttpClientAdapter 使用 dart:io:HttpClient 来发起网络请求。我们看一个 dio 官方提供的例子，代码如下所示：

```
import 'dart:async';
import 'dart:convert';
import 'dart:io';
import 'package:dio/dio.dart';
```

```dart
class MockAdapter extends HttpClientAdapter {
  static const String mockHost = "mockserver";
  static const String mockBase = "http://$mockHost";
  DefaultHttpClientAdapter _defaultHttpClientAdapter =
      DefaultHttpClientAdapter();

  @override
  Future<ResponseBody> fetch(RequestOptions options,
      Stream<List<int>> requestStream, Future cancelFuture) async {
    Uri uri = options.uri;
    if (uri.host == mockHost) {
      switch (uri.path) {
        case "/test":
          return ResponseBody.fromString(
            jsonEncode({
              "errCode": 0,
              "data": {"path": uri.path}
            }),
            200,
            DioHttpHeaders.fromMap({
              HttpHeaders.contentTypeHeader: ContentType.json,
            }),
          );
        case "/download":
          return ResponseBody(
            File("./README.MD").openRead(),
            200,
            DioHttpHeaders.fromMap({
              HttpHeaders.contentLengthHeader: File("./README.MD").lengthSync(),
            }),
          );

        case "/token":
          {
```

```
            var t = "ABCDEFGHIJKLMN".split("")..shuffle();
            return ResponseBody.fromString(
              jsonEncode({
                "errCode": 0,
                "data": {"token": t.join()}
              }),
              200,
              DioHttpHeaders.fromMap({
                HttpHeaders.contentTypeHeader: ContentType.json,
              }),
            );
          }
        default:
          return ResponseBody.fromString("", 404, DioHttpHeaders());
      }
    }
    return _defaultHttpClientAdapter.fetch(
        options, requestStream, cancelFuture);
  }
}
```

该例子自定义了一个 MockAdapter，并且设置了 mockBase 作为 baseUrl，这个 mock 地址一般是本地搭建的（或类似 Rap 等生成的）。当"uri.host==mockHost"时，则进入 switch 条件分支判断逻辑。比如，/test 会进入与其相关的 mock 逻辑，然后通过_defaultHttpClientAdapter.fetch 来执行并返回 Future。最后，我们可以通过"dio.httpClientAdapter= MockAdapter();"的调用方式对 adaptor（适配器）进行切换。

6.4.6　dio 库总结

通过前面的学习，我们了解了 dio 库的基本用法，比如 get/post。然后又学习了 dio 单例、dio 拦截器、dio 拦截器链和 dio 适配器相关的知识。dio 库是一个非常强大的网络请求库，它还支持 FormData 发送数据和多文件上传功能，也支持文件下载、请求认证、证书校验等功能。希望读者在做实际项目时，结合 dio 库官方的相关文档来操作。

6.5 异步编程

说到网络，就一定会提到异步编程。对于涉及网络的操作，在客户端的开发中都是通过异步实现的。在 Flutter 里，异步是用 Future 来修饰的，并运行在 event loop 里。

Flutter 的异步特性和 Android 的 Looper 以及前端的 event loop 有点像，都是不断地从事件队列里获取事件然后运行，并通过异步操作有效防止一些耗时任务（比如网络）对 UI 渲染的影响。我们来看一下 Flutter 是如何做到的。

6.5.1 isolate

Flutter 中很重要的一个概念就是 isolate，它是通过 Flutter Engine 层面的一个线程来实现的，而实现 isolate 的线程又是由 Flutter 管理和创建的。除了 isolate 所在的线程以外，还有其他的线程，它们跟 Flutter 的线程模型（Threading Mode）有关。在我们介绍完 isolate 的基本用法和概念之后，本章后面小节会介绍 Flutter Engine 层线程模型相关的知识。

所有的 Dart 代码都是在 isolate 上运行的。通常情况下，我们的应用都是运作在 main isolate 中的，在必要时我们可以通过相关的 API 创建新的 isolate，以便更好地利用 CPU 的特性，并以此提高效率。需要注意一点，多个 isolate 无法共享内存，必须通过相关的 API 通信才可以。

6.5.2 event loop

另一个比较重要的概念是 event loop。学习过前端的同学一定对 event loop 有所了解，理解它并不困难，而且 Flutter 更简单一些。通过图 6.5 我们来分析一下它的运作原理。

第 6 章 使用网络技术与异步编程

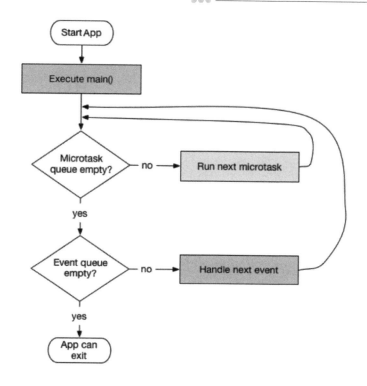

图 6.5

（1）运行 App 并执行 main 方法。

（2）开始并优先处理 microtask queue，直到队列里为空。

（3）当 microtask queue 为空后，开始处理 event queue。如果 event queue 里面有 event，则执行，每执行一条再判断此时新的 microtask queue 是否为空，并且每一次只取出一个来执行。可以这样理解，在处理所有 event 之前我们会做一些事情，并且会把这些事情放在 microtask queue 中。

（4）microtask queue 和 event queue 都为空，则 App 可以正常退出。

注意事项：

当处理 microtask queue 时，event queue 是会被阻塞的。所以 microtask queue 中应避免任务太多或长时间处理，否则将导致 App 的绘制和交互等行为被卡住。所以，绘制和交互等应该作为 event 存放在 event queue 中，这样更合适。

我们来看一个例子，代码如下所示：

```
import 'dart:async';

void main() {
  print('main #1 of 2');
  scheduleMicrotask(() => print('microtask #1 of 3'));

  new Future.delayed(
      new Duration(seconds: 1), () => print('future #1 delayed'));

  new Future(() => print('future #2 of 4'))
      .then((_) => print('future #2a'))
      .then((_) {
        print('future #2b');
        scheduleMicrotask(() => print('microtask #0 from future #2b'));
      })
      .then((_) => print('future #2c'))
      .then((_) => print('future #2d'));

  scheduleMicrotask(() => print('microtask #2 of 3'));

  new Future(() => print('future #3 of 4'))
      .then((_) => new Future(() => print('future #3a a new future')))
      .then((_) => print('future #3b'))
      .then((_) => print('future #3c'));

  new Future(() => print('future #4 of 4'))
      .then((_) {
        new Future(() => print('future #4a'));
      })
      .then((_) => print('future #4b'))
      .then((_) => print('future #4c'));

  scheduleMicrotask(() => print('microtask #3 of 3'));

  print('main #2 of 2');
}
```

在本例中，按照前面描述的 event loop 运行规则，我们在控制台输入 dart main.dart 之后，控制台会展示出以下结果：

```
main #1 of 2
main #2 of 2
microtask #1 of 3
microtask #2 of 3
microtask #3 of 3
future #2 of 4
future #2a
future #2b
future #2c
future #2d
microtask #0 from future #2b
future #3 of 4
future #4 of 4
future #4b
future #4c
future #3a a new future
future #3b
future #3c
future #4a
future #1 delayed
```

这里有两点需要说明。

（1）Future.delayed 表示延迟执行，在设定的延迟时间到了之后，才会被放在 event loop 队列尾部。

（2）Future.then 里的任务不会加入到 event queue 中，要保证异步任务的执行顺序就一定要用 then。

完整的代码见 chapter6/flutter_isolate/lib/main.dart。

在前面第 2 章语法篇里我们已经介绍过 Future，通过英文单词的意思我们可以了解到是"将来"的意思，代表将来会被执行的代码逻辑。每次创建一个 Future，都在 event loop 里添加一条记录，如图 6.6 所示。

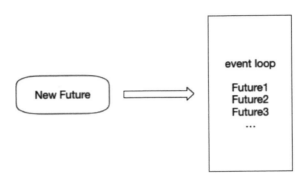

图 6.6

我们可以通过 then 和 whenComplete 方法，完成 Future 包含的操作，并立马执行另一段逻辑。代码如下所示：

```
new Future(() => 21)
    .then((v) => v*2)
    .then((v) => print(v));
```

6.5.3　线程模型与 isolate

前面的小节说过，isolate 是通过 Flutter Engine 层面的一个线程来实现的，而实现 isolate 的线程是由 Flutter 管理和创建的。其实，Flutter Engine 线程的创建和管理是由 embedder（嵌入层）负责的，embeder 是平台引擎移植的中间层代码。我们看一下 Flutter Engine 的运行架构，如图 6.7 所示。

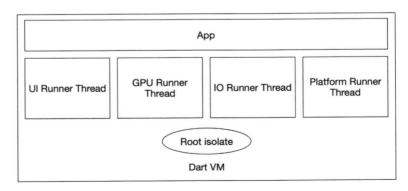

图 6.7

它提供了 4 种 Task Runner。

1. Platform Task Runner

Platform Task Runner 是 Flutter Engine 的主 Task Runner，它不仅可以处理与 Engine 的交互，还可以处理来自 Native（Android/iOS）平台的交互。平台的 API 都只能在主线程中被调用。每一个 Flutter 应用启动的时候都会创建一个 Engine 实例，Engine 创建的时候都会创建一个 Platform Thread 供 Platform Task Runner 使用。即使 Platform Thread 被阻塞，也不会直接导致 Flutter 应用的卡顿。尽管如此，也不建议在这个 Runner 里执行如此繁重的操作，长时间卡住 Platform Thread，应用有可能会被系统的 Watchdog 强行中止。

2. UI Task Runner

UI Task Runner 不能想当然理解为像 Android 那样是运行在主线程的，它其实是运行在线程对应到平台的线程上的，属于子线程。Root isolate 在引擎启动时绑定了不少 Flutter 所需要的方法，从而使其具备调度/提交/渲染帧的能力。在 Root isolate 通知 Flutter Engine 有帧需要被渲染后，渲染时就会生成 Layer Tree 并交给 Flutter Engine。此时，仅生成了需要描绘的内容，然后才创建和更新 Layer Tree，该 Tree 最终决定什么内容会在屏幕上被绘制，因此 UI Task Runner 如果过载会导致卡顿。

isolate 可以理解为单线程，如果运算量大，可以考虑采用独立的 isolate，单独创建的 isolate（非 Root isolate）不支持绑定 Flutter 的功能，也不能调用，只能做数据运算。单独创建的 isolate 的生命周期会受 Root isolate 控制，只要 Root isolate 停止了，那么其他的 isolate 也会停止。isolate 运行的线程是 DartVM 里的线程池提供的。

UI Task Runner 还可以处理来自 Native Plugins 的消息、timers、microtasks、异步 I/O 操作。

3. GPU Task Runner

GPU Task Runner 被用于执行与设备 GPU 相关的调用，它可以将 UI Task Runner 生成的 Layer Tree 所提供的信息转化为实际的 GPU 指令。

GPU Task Runner 的运行线程对应着平台的子线程。UI Task Runner 和 GPU Task Runner 在不同的线程上运行。GPU Runner 会根据目前帧被执行的进度向 UI Task Runner 要求下一帧的数据。在任务繁重的时候，UI Task Runner 会延迟任务进程。这种调度机制确保了 GPU Task Runner 不至于出现过载现象，同时避免了 UI Task Runner 不必要的消耗。如果在该线程中耗时太久的话，就会造成 Flutter 应用的卡顿，因此在 GPU Task Runner 中，尽量不要做耗时的任务。例如，加载图片的同时读取图片数据，就不应该放在 GPU Task Runner 中进行，而应该放在 IO Task Runner 中进行。

4．IO Task Runner

IO Task Runner 的运行线程也对应着平台的子线程。当 UI 和 GPU Task Runner 都出现过载的情况时，会导致 Flutter 应用卡顿。IO Task Runner 就会负责做一些预先处理的读取操作，然后再上报给 GPU Task Runner，且只有 GPU Task Runner 可以访问到 GPU。简单的比喻就是，IO Task Runner 是 GPU Task Runner 的助手，可以减少 GPU Task Runner 的额外操作。IO Task Runner 并不会阻塞 Flutter。虽然在通过 IO Task Runner 加载图片和资源的时候可能会有延迟，但是还是建议为 IO Task Runner 单独建立一个线程。

6.5.4　创建单独的 isolate

前面讲到的 isolate 其实都是 Root isolate，它运行在 UI Task Runner 上。在上一节中，我们也说过，在 UI Task Runner 过载的情况下，可以创建单独的 isolate。单独创建的 isolate 之间没有共享内存，所以它们之间的唯一通信方式只能通过 Port 进行，而且 Dart 中的消息传递总是异步的。isoloate 与线程有着本质的区别，操作系统内的线程之间是可以有共享内存的，而 isolate 不会，这是最为关键的区别。如何创建单独的 isolate 呢？我们来看一个例子，代码如下所示：

```
import 'dart:async';
import 'dart:isolate';
```

```dart
main() async {
  // isolate所需的参数，必须要有SendPort，SendPort需要ReceivePort来创建
  final receivePort = new ReceivePort();
  // 开始创建isolate, Isolate.spawn函数是isolate.dart里的代码，_isolate是我们自己实现的函数
  await Isolate.spawn(_isolate, receivePort.sendPort);
  // 发送的第一个message，是它的SendPort
  var sendPort = await receivePort.first;
  var msg = await sendReceive(sendPort, "foo");
  print('received $msg');
  msg = await sendReceive(sendPort, "bar");
  print('received $msg');
}

/// 新isolate的入口函数
_isolate(SendPort replyTo) async {
  // 实例化一个ReceivePort以接收消息
  var port = new ReceivePort();

  // 把它的sendPort发送给宿主isolate，以便宿主可以给它发送消息
  replyTo.send(port.sendPort);

  // 监听消息，从Port里获取
  await for (var msg in port) {
    var data = msg[0];
    SendPort replyTo = msg[1];
    replyTo.send('应答：' + data);
    if (data == "bar") port.close();
  }
}

/// 对某一个Port发送消息，并接收结果
Future sendReceive(SendPort port, msg) {
  ReceivePort response = new ReceivePort();
```

```
    port.send([msg, response.sendPort]);
    return response.first;
}
```

在 event loop 里存在过度耗时的任务（Task），UI 的操作也被阻塞时，我们可以考虑单独创建一个 isolate。

完整的代码见 chapter6/flutter_isolate/lib/myisolate.dart。

6.5.5 Stream 事件流

Stream 是与 Flutter 相关的一个重要概念，它首先是基于事件流来驱动并设计代码的，然后监听和订阅相关事件，并且对事件的变化进行处理响应。Stream 不是 Flutter 中特有的，而是 Dart 中自带的。我们可以把 Stream 想象成是管道（pipe）的两端，它只允许从一端插入数据并通过管道从另外一端流出数据。我们可以通过 StreamController 来控制 Stream（事件源），比如 StreamController 提供了类型为 StreamSink、属性为 Sink 的控制器作为入口。

Stream 可以传输什么？它支持任何类型数据的传输，包括基本值、事件、对象、集合等，即任何可能改变的数据都可以被 Stream 传递和触发。当我们在传输数据时，可以通过 listen 函数监听来自 StreamController 的属性，在监听之后，可以通过 StreamSubscription 订阅对象并接收 Stream 发送数据变更的通知。

Stream 也是异步处理的核心 API。那么，它与同为异步处理的 Future 有何区别呢？答案是 Future 表示"将来"一次异步获取得到的数据，而 Stream 是多次异步获取得到的数据；Future 将返回一个值，而 Stream 将返回多次值。在讲 Future 时我们提到过 FutureBuilder 类，对于 Stream，它有 StreamBuilder 类负责监听 Stream。当 Stream 数据流出时会自动重新构建组件，并通过 Builder 进行回调。图 6.8 形象地表示手机从云端不断多次获取数据的情况。

第 6 章 使用网络技术与异步编程

图 6.8

我们根据上面提到的 Stream 的相关知识，编写一个简单的例子，部分代码如下所示：

```
class _MyHomePageState extends State<MyHomePage> {
  final StreamController<int> _streamController =
StreamController<int>();
  int _counter = 0;

  @override
  void dispose() {
    _streamController.close();
    super.dispose();
  }

  @override
  Widget build(BuildContext context) {
    return Scaffold(
      appBar: AppBar(
        title: Text(widget.title),
      ),
      body: Center(
        child: Column(
```

```
                mainAxisAlignment: MainAxisAlignment.center,
                children: <Widget>[
                  Text(
                    'You have pushed the button this many times:',
                  ),
                  StreamBuilder<int>(
                      stream: _streamController.stream,
                      initialData: 0,
                      builder: (BuildContext context,
AsyncSnapshot<int> snapshot) {
                        return Text(
                          '${snapshot.data}',
                          style: Theme.of(context).textTheme.display1,
                        );
                      }),
                  Text(
                    '$_counter',
                    style: Theme.of(context).textTheme.display1,
                  ),
                ],
              ),
            ),
            floatingActionButton: FloatingActionButton(
              onPressed: () {
                _streamController.sink.add(++_counter);
              },
              tooltip: 'Increment',
              child: Icon(Icons.add),
            ),
          );
        }
      }
```

这个例子比较简单，就是一个简单的计数器，只不过我们用到了 Stream 相关的知识，把本小节讲述的知识点贯穿起来，即通过 streamController.sink.add 把当前的整型 counter 值传入，在 StreamBuilder 里面去监听 streamController.stream 发生的变化。

完整的代码见 chapter6/flutter_stream。

本章小结

本章我们学习了 HTTP 的相关知识，理解了其基本工作原理，以及 Flutter 如何请求网络。其中，具体讲解了 HttpClient、http、dio 这 3 个库的用法，还包括 JSON 解析、异步编程、线程模型和 isolate 的相关知识。

第 7 章 路　　由

在前文的讲解中，我们已经在一些例子里面，接触过少量跟路由相关的代码了。本章就让我们正式系统地学习与路由相关的知识。

7.1　路由简介

路由是连接界面的"桥梁"，这个"桥梁"就是 Navigator，即导航的意思。导航用于管理一组具有某种进出规则的页面的组件，以此来实现各个页面之间有规律的切换。维持这种规律并且存放路由信息的事物我们称之为"路由栈"。

路由的概念在整个大前端知识体系中已经很成熟了，Native 和 Web 前端中都有涉及。在 Native 中，路由对组件化工程进行了解耦，比如阿里巴巴的 ARoute。路由在 Web 中的单页应用中也有涉及，被用来管理页与页之间的跳转和参数传递。带路由设计的前端框架包括 Vue 和 React。

Flutter 中的路由借鉴了一些前端的思想。下面让我们看一下路由的基本用法。

7.1.1 基本用法

简单来说，路由跳转的关系如图 7.1 所示。

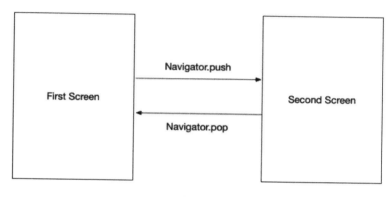

图 7.1

在使用方法上，一个界面跳转到另一个界面使用 Navigator.push 方法，而从一个界面返回到上一个界面一般用 Navigator.pop 方法。

当然，在实际项目中，情况没那么简单，页面之间的跳转情况比较多，这就涉及路由栈管理的一些知识了，后面的小节笔者会介绍。

路由的种类分成静态路由和动态路由，下面，我们分别对这两种情况进行讲解。

7.1.2 静态路由

静态路由在明确知道要往哪里跳转的情况下使用。在 MaterialApp 构造函数里，我们可以定义路由列表。通过前面几章的学习，我们知道，main 方法通常是一个程序的入口函数，也是程序运行的最高"统治者"。在我们创建完成 Flutter 项目时，会通过 runApp 方法传入第一个，也是最底部的实例，类似代码如下所示：

```
void main() => runApp(MyApp());
```

这里，我们传入的是 MyApp，其代码结构大致如下所示：

```
class MyApp extends StatelessWidget {

  @override
  Widget build(BuildContext context) {
    return MaterialApp(
      title: 'Flutter路由',
      theme: ThemeData(
        primarySwatch: Colors.blue,
      ),
      routes: {
        '/page1': (context) => Page1(),
        '/page2': (context) => Page2(),
        '/page3': (context) => Page3(),
        '/page4': (context) => Page4(),
      },
      onUnknownRoute: (RouteSettings setting) {
        String name = setting.name;
        print("未匹配到路由:$name");
        return new MaterialPageRoute(builder: (context) {
          return new ErrorPage(title: name);
        });
      },
      home: Page1(),
    );
  }
}
```

在这段代码中，我们定义了 home 的第一屏，即代码中的 Page1 是"路由栈"中的第一个且是最底部的实例。要往路由栈里加东西，我们可以通过调用路由的一些方法来实现，比如 Navigator.push。

在 routes 配置项里，我们定义了路由表（也称为命名路由），page1、page2、page3、page4 都是我们定义的静态路由。routes 的本质就是 Map。另外需要注意的是，MaterialApp 还有另一个属性，它就是 initialRoute。它指的是 App 启动时默认的路由，如果使用了该属性，则不需要定义 home 属性。上面定义的是路由表，该表只是告诉 Flutter 怎样通过路由表定义的名称去

· 198 ·

找到对应的界面。根据官方的建议，我们可以通过 Navigator.pushNamed 方法进行跳转。

在实际项目中，考虑到程序的健壮性，我们也可以自定义一个路由出错界面，可以通过 onUnknownRoute 来实现。在上述代码中，当跳转到一个并不存在的路由时就跳转到了一个自定义的 ErrorPage 中。我们看一下 Navigator.pushNamed 是怎样使用的，代码如下所示：

```
class Page1 extends StatelessWidget {
  final String title;

  Page1({Key key, @required this.title}) : super(key: key);

  @override
  Widget build(BuildContext context) {
    return Scaffold(
      appBar: AppBar(
        title: Text(this.title),
      ),
      body: Center(
        child: RaisedButton(
          child: Text('跳转到 Page2'),
          onPressed: () {
            Navigator.pushNamed(context, '/page2');
          },
        ),
      ),
    );
  }
}
```

在 RaisedButton 的 onPressed 里，我们调用方法 Navigator.pushNamed (context, 'page2')，这样就可以跳转到 page2 界面。如果用 Navigator.push，则可以这样写，代码如下所示：

```
Navigator.of(context).push(
  new MaterialPageRoute(
    builder: (context) {
```

```
        return new Page2();
     },
   ),
);
```

相比之下，我们采用路由表的方式，通过 Navigator.pushNamed 来操作路由的过程会更简洁。

完整的代码见 chapter7/flutter_routes/lib/page1.dart。

7.1.3 动态路由

前面我们学习了静态路由，然而，在大多数场景下，我们在两个界面之间跳转时需要携带参数，比如从列表界面到详情界面这种情况。这时候，动态路由就起作用了。下面我们来探索一下动态路由是怎样进行参数传递的。我们看一下 Flutter 官方的 Todo List 例子，创建步骤如下所示。

（1）创建一个 Todo 的实体类，代码如下所示：

```
class Todo {
  final String title;
  final String description;

  Todo(this.title, this.description);
}
```

（2）创建并生成一个 Todo 的 List 列表，代码如下所示：

```
final todos = List<Todo>.generate(
  20,
  (i) => Todo(
    'Todo $i',
    'A description of what needs to be done for Todo $i',
  ),
);
```

（3）通过 ListView 进行数据展示，代码如下所示：

```
ListView.builder(
  itemCount: todos.length,
  itemBuilder: (context, index) {
    return ListTile(
      title: Text(todos[index].title),
    );
  },
);
```

（4）创建 Todo List 的详情页面进行数据展示。在详情页面的展示中，Todo 的 title 字段和 description 分别作为标题和内容，其中 Todo 对象是从列表传递过来的。代码如下所示：

```
class DetailScreen extends StatelessWidget {
  final Todo todo;

  DetailScreen({Key key, @required this.todo}) : super(key: key);

  @override
  Widget build(BuildContext context) {
    return Scaffold(
      appBar: AppBar(
        title: Text("${todo.title}"),
      ),
      body: Padding(
        padding: EdgeInsets.all(16.0),
        child: Text('${todo.description}'),
      ),
    );
  }
}
```

（5）通过动态路由跳转并传递参数，代码如下所示：

```
ListView.builder(
  itemCount: todos.length,
  itemBuilder: (context, index) {
```

```
    return ListTile(
      title: Text(todos[index].title),

      // 创建详情页的实例并传递 Todo 对象作为参数
      onTap: () {
        Navigator.push(
          context,
          MaterialPageRoute(
            builder: (context) => DetailScreen(todo: todos[index]),
          ),
        );
      },
    );
  },
);
```

最后，运行效果如图 7.2 所示。

Todos
Todo 0
Todo 1
Todo 2
Todo 3
Todo 4
Todo 5
Todo 6
Todo 7
Todo 8
Todo 9

图 7.2

我们通过点击列表产生点击事件，这个时候调用的就是动态路由的跳转方法，代码如下所示：

```
Navigator.push(
  context,
  MaterialPageRoute(
    builder: (context) => DetailScreen(todo: todos[index]),
  ),
);
```

在这个方法中，我们传入了 DetailScreen 和 Todo 实体类作为参数，从而实现了界面跳转。最后，展现的 DetailScreen 界面如图 7.3 所示。

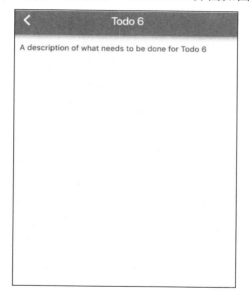

图 7.3

完整的代码见 chapter7/flutter_routes/lib/todo.dart。

7.1.4 参数回传

在很多业务场景下，我们需要从 A 界面跳转到 B 界面，在 B 界面进行一些操作后，再把操作结果返回 A 界面。比如应用中的城市选择就是一个很典型的例子。在这种业务场景下，我们就需要使用带参数返回的路由。

下面，我们选择一个学生选课的例子作为应用场景，来讲解一下路由参数是怎样回传的，完整例子的代码如下所示：

```dart
class HomeScreen extends StatelessWidget {

  @override
  Widget build(BuildContext context) {
    return Scaffold(
      appBar: AppBar(
        title: Text('参数回传'),
      ),
      body: Center(child: SelectionButton()),
    );
  }
}

class SelectionButton extends StatelessWidget {
  @override
  Widget build(BuildContext context) {
    return RaisedButton(
      onPressed: () {
        _navigateAndDisplaySelection(context);
      },
      child: Text('课程选择'),
    );
  }

  _navigateAndDisplaySelection(BuildContext context) async {
    final result = await Navigator.push(
      context,
      MaterialPageRoute(builder: (context) => SelectionScreen()),
    );

    Scaffold.of(context)
      ..removeCurrentSnackBar()
      ..showSnackBar(SnackBar(content: Text("$result")));
  }
```

```
}

class SelectionScreen extends StatelessWidget {

  @override
  Widget build(BuildContext context) {
    return Scaffold(
      appBar: AppBar(
        title: Text('请选择一门课程'),
      ),
      body: Center(
        child: Column(
          mainAxisAlignment: MainAxisAlignment.center,
          children: <Widget>[
            Padding(
              padding: const EdgeInsets.all(8.0),
              child: RaisedButton(
                onPressed: (){
                  Navigator.pop(context, 'Android');
                },
                child: Text('Android'),
              ),
            ),
            Padding(
              padding: const EdgeInsets.all(8.0),
              child: RaisedButton(
                onPressed: (){
                  Navigator.pop(context, 'IOS');
                },
                child: Text('IOS'),
              ),
            ),
            Padding(
              padding: const EdgeInsets.all(8.0),
              child: RaisedButton(
                onPressed: (){
                  Navigator.pop(context, 'Flutter');
                },
```

```
                    child: Text('Flutter'),
                ),
            )
        ],
        ),
    ),
    );
  }
}
```

在本例中，与之前介绍的动态路由跳转方式类似，只不过是通过 await 的方式，给 Navigator 的 push 结果添加了一个 result，这个 result 就是回传的结果，代码如下所示：

```
final result = await Navigator.push(
  context,
  MaterialPageRoute(builder: (context) => SelectionScreen()),
);
```

至于怎样处理 result 回传结果，可以根据实际业务场景的需要来处理。在本例中，我们通过 SnackBar 把选择结果回调显示出来。需要注意的是，pop 方法需要在路由出栈时把参数一起返回来，代码如下所示：

```
Navigator.pop(context, 'Flutter');
```

完整的代码见 chapter7/flutter_route/lib/navigator_with_result.dart。

7.2 路由栈

在前文中，我们接触了一些路由的控制方法，也提到过"路由栈"的基本概念。但实际上，路由的控制方法不止这些，这是因为实际情况更加复杂多变。那么，本节我们来更加深入地了解路由栈和学习更多的路由控制方法。

7.2.1 路由栈详解

我们看一下，一个路由通过 push（pushNamed 也一样）方法添加之后，其路由栈如图 7.4 所示。

当我们调用 pop 方法出栈后，路由栈又把 page2 移除了，此时，路由栈如图 7.5 所示。

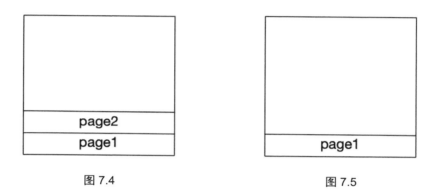

图 7.4　　　　　　　　　　　图 7.5

因此，我们可以得出一个结论：Navigator 里的 push 方法把元素添加到堆栈的顶部，而 pop 方法则删除了顶部的元素。

上述描述的入栈与出栈只是简单的操作。

在 Android 里面，Android 的启动模式有 4 种，即 standard、singleTop、singleTask、singleInstance，通过 intent 跳转之后，它们的 Activity 栈是不一样的。前文介绍的 push、pop 与 Android 里的 standard 启动模式类似。Flutter 也考虑到了类似 Android 的各种启动模式的实现，我们来看一下 Flutter 是如何做到的。

7.2.2　pushReplacementNamed 方法

假设现在的路由栈里有 3 个 page，其路由栈情况是下面这样的，如图 7.6 所示。

然后，再假设我们需要跳转到一个 page4 界面。这一次，我们不通过 page3 直接 push，而是调用另一个方法 pushReplacementNamed，代码如下所示：

```
Navigator.of(context).pushReplacementNamed('/page4');
```

这个时候，路由栈的情况，如图 7.7 所示。

page3		page4
page2		page2
page1		page1

图 7.6　　　　　　　　图 7.7

由于我们调用了 pushReplacementNamed（其实就是替换的意思）方法，执行完 pushReplacementNamed 方法后，page3 就没有了。如果这个时候执行 pop 方法，则路由栈里就只剩下 page1 和 page2 了，如图 7.8 所示。

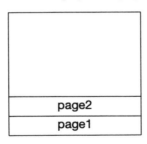

图 7.8

7.2.3　popAndPushNamed 方法

与 pushReplacementNamed 方法相似的有 popAndPushNamed 方法，其执行结果展示的路由栈与 pushReplacementNamed 方法的效果是一致的。

不同的是,page3 会同时有 pop 的转场效果和从 page2 页 push 的转场效果(即调用了 popAndPushNamed 之后,同时有弹出和推进动画的转场效果)。从交互体验来讲,popAndPushNamed 有 pop 的效果,是一种选择并携带选择结果返回的效果。

7.2.4　pushNamedAndRemoveUntil 方法

在 App 里,有一个普遍存在的场景,即打开一个 App 之后,会出现 App 的启动页,然后进入欢迎页(或引导页),最后才进入 App 的首页。在这种情况下,用户选择返回,是应该从首页退出 App 的,而不是再次倒退到欢迎页和启动页。这个时候,pushNamedAndRemoveUntil 方法就派上用场了。我们可以通过以下的方式调用,让整个路由栈里只存在一个界面,调用代码如下所示:

```
Navigator.of(context).pushNamedAndRemoveUntil(
    '/homepage', (Route <dynamic> route) => false);
```

其中,"(Route <dynamic> route) => false)" 能确保删除先前所有路由栈中的页面实例。最后,路由栈中的结果如图 7.9 所示。

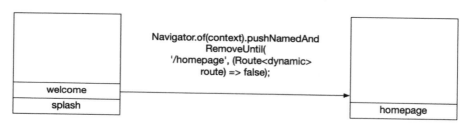

图 7.9

有的时候,我们只指定几个路由,比如从 page1 依次跳转到 page5,在 pop 时希望从 page5 直接回到 page2,这个时候应该怎么做呢?我们可以改装一下 pushNamedAndRemoveUntil 方法的参数,执行代码如下所示:

```
Navigator.of(context).pushNamedAndRemoveUntil('/page5',
ModalRoute.withName('/page2'));
```

此时，路由栈的变化如图 7.10 所示。

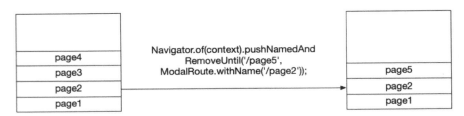

图 7.10

7.2.5 popUntil 方法

popUntil 和上面说的 pushNamedAndRemoveUntil 的第二种用法有点类似，只不过没有 push 操作，而是直接 pop 到指定界面，我们来看一下调用方式，代码如下所示：

```
Navigator.popUntil(context, ModalRoute.withName('/page2'));
```

此时，路由栈的变化如图 7.11 所示。

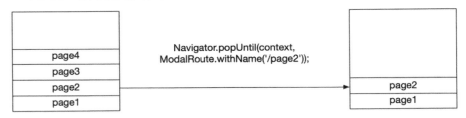

图 7.11

7.3 自定义路由

我们创建的所有的路由都是 MaterialPageRoute。有的时候我们需要改变默认的路由转场效果，这个时候就需要做一些定制，要用到另一个类了，它就是 PageRouteBuilder。我们先看一下 PageRouteBuilder 对应的源码，其构造函数如下所示：

```
PageRouteBuilder({
  RouteSettings settings,
  @required this.pageBuilder,
  this.transitionsBuilder = _defaultTransitionsBuilder,
  this.transitionDuration = const Duration(milliseconds: 300),
  this.opaque = true,
  this.barrierDismissible = false,
  this.barrierColor,
  this.barrierLabel,
  this.maintainState = true,
}) : assert(pageBuilder != null),
     assert(transitionsBuilder != null),
     assert(barrierDismissible != null),
     assert(maintainState != null),
     assert(opaque != null),
     super(settings: settings);
```

我们看一下 PageRouteBuilder 中几个重要的属性，如表 7.1 所示。

表 7.1

属性	取值
Opaque	是否遮挡整个屏幕
transitionsBuilder	用于自定义的转场效果
pageBuilder	用来创建所要跳转到的页面
transitionDuration	转场动画的持续时间

在第 4 章中，我们学习了与动画相关的知识，下面举一个完整的例子。通过点击一个自定义的 Widget 和 PageRouteBuilder，实现一个简单的 Hero 效果的路由切换。首先定义一个通用方法，用于路由切换，代码如下所示：

```
_onButtonTapCustom(Widget page) {
  Navigator.of(context).push(
    PageRouteBuilder<Null>(
      pageBuilder:(BuildContext context, Animation<double> animation,
        Animation<double> secondaryAnimation) {
        return AnimatedBuilder(
          animation: animation,
          builder: (BuildContext context, Widget child) {
```

```
            return Opacity(
              opacity: animation.value,
              child: page,
            );
          });
    },
    transitionDuration: Duration(milliseconds: 600)),
  );
}
```

然后，可以定义我们需要通过Hero动画变化的一个CustomLogo，代码如下所示：

```
class CustomLogo extends StatelessWidget {
  final double size;

  CustomLogo({this.size = 200.0});

  @override
  Widget build(BuildContext context) {
    return Container(
      color: Colors.lightBlueAccent,
      width: size,
      height: size,
      child: Center(
        child: FlutterLogo(
          size: size,
        ),
      ),
    );
  }
}
```

接下来，我们定义 Hero 的源和目标，代码如下所示：

```
body: Center(
    child: Column(
      children: <Widget>[
        RaisedButton(
```

```
            padding: EdgeInsets.all(10.0),
            onPressed: () {
              _onButtonTapCustom(PageRouteBuilderResult());
            },
            child: Text(
              '调用自定义路由',
              textAlign: TextAlign.center,
              style: TextStyle(fontSize: 13.0),
            ),
          ),
          Hero(
            tag: "hero1",
            child: ClipOval(
              child: CustomLogo(
                size: 60.0,
              ),
            ),
          ),
          Hero(
            tag: "hero2",
            child: Material(
              color: Colors.transparent,
              child: Text(
                "Hero效果",
                style: TextStyle(fontSize: 14.0, color: Colors.black),
              ),
            ))
        ],
    )),
```

上面我们定义了两个 Hero 动画的 tag，根据 Hero 的约定规则，目标页面的 Widget 也需要指定 Hero 的 tag，部分代码如下所示：

```
Align(
  alignment: Alignment.center,
  child: Hero(
    tag: "hero1",
    child: Container(
```

```
          height: 250.0,
          width: 250.0,
          child: CustomLogo(),
        ),
      ),
    ),
```

最后，运行本例，效果如图 7.12 所示。

点击对应按钮，调用我们自定义的 PageRouteBuilder 路由，再加上 Hero 动画的变换，效果如图 7.13 所示。

图 7.12

图 7.13

完整的代码见 chapter7/flutter_route/lib/testpageroutebuilder.dart。

本章小结

本章学习了路由的几种跳转方式，理解了 Navigator 是负责管理路由的。在路由栈里，可以通过 Navigator.push 或 Navigator.pushNamed 进行入栈操作，可以通过 Navigator.pop 进行出栈操作。同时，还介绍了静态和动态两类路由、路由表和参数回传的用法、路由栈的概念，以及一些其他管理路由的方法，比如 popAndPushNamed、popUntil 等方法。

第 8 章 持 久 化

在本章中，我将为大家介绍 Flutter 的持久化方式。相信，对于有过 App 开发经验的人来说，这个话题不会陌生。比如，在 Android 里面的 SharedPreferences、数据库 sqlite、文件读写，还包括在服务器上的存储以及 HTML5 的 localStorage、PWA。合理地运用持久化，可以让 App 支持离线化，给用户体验带来极大的提升。下面，我们就来讲解一下 Flutter 中的一些持久化方式。

8.1 shared_preferences本地存储

Flutter 官方推荐我们使用 shared_preferences 进行数据存储，它是 Flutter 社区开发的一个本地数据存取插件，具有以下特性。

（1）它是一个异步的、简单的、持久化的、key-value 形式的存储系统。

（2）在 Android 端，它是基于 SharedPreferences 开发的。

（3）在 iOS 端，它是基于 NSUserDefaults 开发的。

如何使用它呢？

在 pubspec.yaml 里添加以下依赖就可以实现，代码如下所示：

```
dependencies:
  shared_preferences: ^0.5.1+1
```

8.1.1 shared_preferences 的常用操作

shared_preferences 的常用操作比较简单，我们直接看一下以下的常用方法。

保存的方式，代码如下所示：

```
SharedPreferences prefs = await SharedPreferences.getInstance();
prefs.setString(key, value);
```

读取的方式，代码如下所示：

```
SharedPreferences prefs = await SharedPreferences.getInstance();
prefs.getString(key);
```

删除的方式，代码如下所示：

```
SharedPreferences prefs = await SharedPreferences.getInstance();
prefs.remove(key);
```

8.1.2 shared_preferences 举例

在本例中，我们改造一下项目生成时默认的计数器，代码如下所示：

```
import 'package:flutter/material.dart';
import 'package:shared_preferences/shared_preferences.dart';

void main() => runApp(MyApp());

class MyApp extends StatelessWidget {
  @override
  Widget build(BuildContext context) {
    return MaterialApp(
      title: 'Flutter Demo',
      theme: ThemeData(
        primarySwatch: Colors.blue,
```

```dart
    ),
      home: MyHomePage(title: 'Flutter Demo Home Page'),
    );
  }
}

class MyHomePage extends StatefulWidget {
  MyHomePage({Key key, this.title}) : super(key: key);

  final String title;

  @override
  _MyHomePageState createState() => _MyHomePageState();
}

class _MyHomePageState extends State<MyHomePage> {

  int _count = 0;

  @override
  void initState() {
    super.initState();
    _getCounter();
  }

  void _incrementCounter() async {
    SharedPreferences prefs = await SharedPreferences.getInstance();
    int counter = (prefs.getInt("counter") ?? 0) + 1;
    setState(() {
      _count = counter;
    });
    await prefs.setInt('counter', counter);
  }

  _getCounter() async {
    SharedPreferences prefs = await SharedPreferences.getInstance();
    setState(() {
      _count = prefs.get('counter') ?? 0;
```

```dart
      });
    }

    @override
    Widget build(BuildContext context) {
      return Scaffold(
        appBar: AppBar(
          title: Text(widget.title),
        ),
        body: Center(
          child: Column(
            mainAxisAlignment: MainAxisAlignment.center,
            children: <Widget>[
              Text(
                '点击数: ' + _count.toString(),
                style: Theme.of(context).textTheme.display1,
              ),
            ],
          ),
        ),
        floatingActionButton: FloatingActionButton(
          onPressed: _incrementCounter,
          tooltip: 'Increment',
          child: Icon(Icons.add),
        ),
      );
    }
  }
```

看一下效果，如图 8.1 所示。

在本例中，我们为_count 计数器变量进行了保存和读取，其操作访问的 API 就是 shared_preferences 的 API。在每次被点击时，为计数器增加值的方法就是_incrementCounter。另外，在每次进入应用时程序会调用_getCounter 方法，这样就保证了点击的次数被持久化地保存在 shared_preferences 里面。

完整的代码见 chapter8/flutter_store/lib/share_preferences.dart。

图 8.1

8.2 SQLite数据库

数据库最早是在服务端运用的，后来出现了 Android 和 iOS 系统，在手机端，数据库也开始被使用，数据库也是常用的持久化方式之一。说到数据库，这里介绍一下 SQLite，它是在手机端上最老牌、最流行的数据库。使用起来比 shared_preferences 稍微复杂一些。前面介绍的 shared_preferences 我们也可以理解为 key-value 存储模式，在需要对数据进行大批量地增、删、改、查（GRUD）操作时，我们就会使用数据库。举一个例子，在一个应用中，有时需要在离线模式下也保存一些信息以便给用户更好的体验，比如新闻列表，这时候就用到了本地数据库存储。Flutter 为我们提供了 sqflite 插件来进行 GRUD 操作，下面我们来一起学习 sqflite 的使用。

8.2.1 sqflite 依赖库简介

目前与数据库相关，并且使用最广的依赖库就是 sqflite。它同时支持 Android 和 iOS 版本。我们可以在 Flutter Packages 上搜索 sqflite，找到下载地址和使用说明。sqflite 有以下特性。

（1）支持事物和批处理。

（2）支持自动 Version 管理。

（3）支持增、删、改、查的 Helpers 工具类。

（4）支持 Android/iOS 后台线程的运行。

下面，我们来看一下怎样引入 sqflite，代码如下所示：

```
dependencies:
  sqflite: ^1.1.3
```

sqflite 常用的操作方式有 8 种，我们依次来看一下。

（1）获取和删除 database，代码如下所示：

```
var databasesPath = await getDatabasesPath();
String path = join(databasesPath, 'demo.db');
await deleteDatabase(path);
```

（2）打开 database，在建立 db 的同时创建数据库，代码如下所示：

```
Database database = await openDatabase(path, version: 1,
    onCreate: (Database db, int version) async {
  await db.execute(
      'CREATE TABLE Test (id INTEGER PRIMARY KEY, name TEXT, value INTEGER, num REAL)');
});
```

（3）插入数据有两种方式，第一种方式的代码如下所示：

```
Future<int> rawInsert(String sql, [List<dynamic> arguments]);
```

rawInsert 方法的第一个参数为一条插入的 sql 语句，使用"?"作为占位符，然后通过 List 的参数传入填充数据。例如以下的插入语句：

```
int id = await txn.rawInsert(
    'INSERT INTO Test(name, value, num) VALUES(?, ?, ?)',
    ['another name', 12345678, 3.1416]);
print('inserted2: $id');
```

需要注意的是，参数个数与"?"的个数一致，上面代码运行的效果是插入数据后，返回数据库所存入的 id。

第二种方式的代码如下所示：

```
Future<int> insert(String table, Map<String, dynamic> values,
    {String nullColumnHack, ConflictAlgorithm conflictAlgorithm});
```

insert 方法的第一个参数是需要操作的表名，第二个参数是 Map 的类型，用来传入需要添加的字段名和对应的取值。

（4）修改操作有两种方式，第一种方式的代码如下所示：

```
Future<int> rawUpdate(String sql, [List<dynamic> arguments]);
```

和前面的插入操作类似，只不过这里的方法名为 rawUpdate。

第二种方式的代码如下所示：

```
Future<int> update(String table, Map<String, dynamic> values,
    {String where,
    List<dynamic> whereArgs,
    ConflictAlgorithm conflictAlgorithm});
```

update 方法第一个参数是操作的表名，第二个参数是修改的字段和对应值，后边的可选参数依次表示 where 查询子句（可使用"?"作为占位符）、where 子句所对应的占位符参数值以及发生冲突时的操作算法策略（例如回滚、终止、忽略等）。

（5）查询操作有两种方式，第一种方式的代码如下所示：

```
Future<List<Map<String, dynamic>>> rawQuery(String sql, [List
<dynamic> arguments]);
```

rawQuery 方法的第一个参数为查询 sql 语句，可以使用"？"作为占位符，它通过传入第二个参数来填充数据。

第二种方式的代码如下所示：

```
Future<List<Map<String, dynamic>>> query(String table,
    {bool distinct,
    List<String> columns,
    String where,
    List<dynamic> whereArgs,
    String groupBy,
    String having,
    String orderBy,
    int limit,
    int offset});
```

query 方法的第一个参数是操作的表名，后边的可选参数依次表示是否去重、查询字段、where 的查询子句、where 子句的占位符参数值、分组查询（groupBy）子句、having 子句、排序（orderBy）子句、查询上限条数（limit）、查询的偏移位（offset）。

（6）操作删除通常分为两种，即物理删除和逻辑删除。

物理删除指的是操作真实的数据库并实现删除操作，而逻辑删除指的是通过 update 语句修改数据库，并且在查询时过滤掉不符合查询条件的记录，下面我们看一下 sqflite 里怎样实现逻辑删除。

第一种方式的代码如下所示：

```
Future<int> rawDelete(String sql, [List<dynamic> arguments]);
```

和之前的查询和插入语句类似，rawDelete 方法的第一个参数为一条删除 sql 语句，可以使用"？"作为占位符，并通过第二个参数来填充数据。

第二种方式的代码如下所示：

```
Future<int> delete(String table, {String where, List<dynamic> whereArgs});
```

delete 方法的第一个参数是操作的表名，后边的可选参数依次表示 where 子句（可使用"?"作为占位符）、where 子句占位符的参数值。

（7）计算总记录数的代码如下所示：

```
Sqflite.firstIntValue(await database.rawQuery('SELECT COUNT(*) FROM Test'));
```

（8）关闭数据库，数据库对象在使用完之后要在适当的时候关闭，方法如下所示：

```
await database.close();
```

以上的 8 个操作是 sqflite 的基本操作，需要牢固掌握。

8.2.2　封装 SQL Helpers

在实际项目中，我们在操作表的时候，通常不会一条一条地通过 sql 语句对数据库进行操作，而是通过一些封装类，封装成 SQL Helpers 来进行操作。有 Android 项目经验的人一定用过类似 SQL Helpers 的帮助类。下面，我们举一个有关 SQL Helpers 封装类的例子，需要操作的对象是 Todo 数据库表，代码如下所示：

```
final String tableTodo = 'todo';
final String columnId = '_id';
final String columnTitle = 'title';
final String columnDone = 'done';

class Todo {
  int id;
  String title;
  bool done;

  Map<String, dynamic> toMap() {
    var map = <String, dynamic>{
      columnTitle: title,
      columnDone: done == true ? 1 : 0
```

```
    };
    if (id != null) {
      map[columnId] = id;
    }
    return map;
  }

  Todo();

  Todo.fromMap(Map<String, dynamic> map) {
    id = map[columnId];
    title = map[columnTitle];
    done = map[columnDone] == 1;
  }
}

class TodoProvider {
  Database db;

  Future open(String path) async {
    db = await openDatabase(path, version: 1,
        onCreate: (Database db, int version) async {
      await db.execute('''
create table $tableTodo ( 
  $columnId integer primary key autoincrement, 
  $columnTitle text not null, 
  $columnDone integer not null)
''');
    });
  }

  Future<Todo> insert(Todo todo) async {
    todo.id = await db.insert(tableTodo, todo.toMap());
    return todo;
  }

  Future<Todo> getTodo(int id) async {
    List<Map> maps = await db.query(tableTodo,
```

```
      columns: [columnId, columnDone, columnTitle],
      where: '$columnId = ?',
      whereArgs: [id]);
  if (maps.length > 0) {
    return Todo.fromMap(maps.first);
  }
  return null;
}

Future<int> delete(int id) async {
  return await db.delete(tableTodo, where: '$columnId = ?',
whereArgs: [id]);
}

Future<int> update(Todo todo) async {
  return await db.update(tableTodo, todo.toMap(),
      where: '$columnId = ?', whereArgs: [todo.id]);
}

Future close() async => db.close();
}
```

通过 SQL Helpers 封装类，我们可以对数据库进行反复操作。首先实例化 TodoProvider，然后调用对应的 SQL Helpers 里的方法（open、insert、getTodo、delete、update 等）即可，这样也不容易出错。

8.2.3　sqflite 实现员工打卡示例

下面，我们用 sqflite 来实现一个简单的员工打卡系统。大致的需求是用户在登录应用时有一个输入框，可以输入员工姓名，然后点击保存，保存完之后可以在列表记录并显示所有已打卡的用户，打卡完后可以进入打卡浏览界面浏览所有打过卡的员工，也可以在保存界面清除打卡数据。

这一次，我们封装一个单例模式的 DBProvider 类，通过直接调用 DBProvider.db 就可以访问 DBProvider 实例，并且可以直接调用封装的数据库操作方法。代码如下所示：

```dart
import 'dart:io';

import 'package:path/path.dart';
import 'package:path_provider/path_provider.dart';
import 'package:sqflite/sqflite.dart';

final String tableUser = 'user';
final String columnId = '_id';
final String columnName = 'name';

class User {
  int id;
  String name;

  Map<String, dynamic> toMap() {
    var map = <String, dynamic>{
      columnName: name,
    };
    if (id != null) {
      map[columnId] = id;
    }
    return map;
  }

  User();

  User.fromMap(Map<String, dynamic> map) {
    id = map[columnId];
    name = map[columnName];
  }
}

class DBProvider {
  DBProvider._();

  static final DBProvider db = DBProvider._();
```

```
    Database _database;

    Future<Database> get database async {
      if (_database != null) return _database;
      _database = await initDB();
      return _database;
    }

    initDB() async {
      Directory documentsDirectory = await getApplicationDocumentsDirectory();
      String path = join(documentsDirectory.path, "UserDB.db");
      return await openDatabase(path, version: 1, onOpen: (db) {},
          onCreate: (Database db, int version) async {
        await db.execute('''
            create table $tableUser (
              $columnId integer primary key autoincrement,
              $columnName text not null)
            ''');
      });
    }

    Future<User> insert(User user) async {
      final db = await database;
      user.id = await db.insert(tableUser, user.toMap());
      return user;
    }

    Future<User> getUser(int id) async {
      final db = await database;
      List<Map> maps = await db.query(tableUser,
          columns: [columnId, columnName],
          where: '$columnId = ?',
          whereArgs: [id]);
      if (maps.length > 0) {
        return User.fromMap(maps.first);
      }
      return null;
```

```
  }

  Future<List<User>> getAllUser() async {
    final db = await database;
    var res = await db.query("User");
    List<User> list =
        res.isNotEmpty ? res.map((c) => User.fromMap(c)).toList() : [];
    return list;
  }

  Future<int> delete(int id) async {
    final db = await database;
    return await db.delete(tableUser, where: '$columnId = ?',
        whereArgs: [id]);
  }

  Future<int> update(User user) async {
    final db = await database;
    return await db.update(tableUser, user.toMap(),
        where: '$columnId = ?', whereArgs: [user.id]);
  }

  removeAll() async {
    final db = await database;
    db.delete(tableUser);
  }

  Future close() async {
    final db = await database;
    db.close();
  }
}
```

代码运行之后，效果如图 8.2 所示。

然后，我们创建一个界面，可以输入用户名并进行保存，代码如下所示：

第 8 章 持久化

图 8.2

```
import 'package:flutter/material.dart';
import 'package:flutter_store/check_in_result.dart';
import 'package:flutter_store/db_helpers.dart';
import 'package:shared_preferences/shared_preferences.dart';

class CheckInPage extends StatefulWidget {
  CheckInPage({Key key, this.title}) : super(key: key);

  final String title;

  @override
  _CheckInPageState createState() => _CheckInPageState();
}

class _CheckInPageState extends State<CheckInPage> {
  final TextEditingController textTitleEditingController =
      new TextEditingController();

  _save() {
    User user = User();
    user.name = textTitleEditingController.text;
```

```
    DBProvider.db.insert(user);
    Navigator.of(context).push(
      new MaterialPageRoute(
        builder: (context) {
          return new CheckInResult();
        },
      ),
    );
}

_removeData() {
  DBProvider.db.removeAll();
}

@override
Widget build(BuildContext context) {
  return Scaffold(
    appBar: AppBar(
      title: Text(widget.title),
    ),
    body: Center(
        child: Column(
      children: <Widget>[
        TextField(
          controller: textTitleEditingController,
          autofocus: true,
          decoration: new InputDecoration(
            hintText: '请输入用户名',
          ),
        ),
        RaisedButton(
          child: Text('保存'),
          onPressed: _save,
        ),
        RaisedButton(
          child: Text('清除数据'),
```

```
          onPressed: _removeData,
        )
      ],
    )),
  );
 }
}
```

最后，我们在列表界面中查询出所有已打卡的用户，代码如下所示：

```
import 'package:flutter/foundation.dart';
import 'package:flutter/material.dart';
import 'package:flutter_store/db_helpers.dart';

class CheckInResult extends StatelessWidget {
  CheckInResult({Key key}) : super(key: key);

  @override
  Widget build(BuildContext context) {
    return Scaffold(
      appBar: AppBar(title: Text("已打卡人员")),
      body: FutureBuilder<List<User>>(
        future: DBProvider.db.getAllUser(),
        builder:(BuildContext context, AsyncSnapshot<List<User>> snapshot) {
          if (snapshot.hasData) {
            return ListView.builder(
              itemCount: snapshot.data.length,
              itemBuilder: (BuildContext context, int index) {
                User item = snapshot.data[index];
                return ListTile(
                  title: Text(item.name),
                );
              },
            );
          } else {
            return Center(child: CircularProgressIndicator());
          }
```

```
            },
         ),
      );
   }
}
```

代码运行之后，效果如图 8.3 所示。

图 8.3

完整的代码见 chapter8/flutter_store/lib/sqflite_checkin.dart。

8.3 文件形式存储

前面我们已经讲解了基于 key-value 形式的 SharedPreferences 存储方式和 sqflite 的数据库插件使用方式。对于持久化知识，我们还需要对文件形式存储有一定的了解。在 Flutter 中已经实现了文件操作（File）相关的 API。Flutter 中使用 File 获取手机中的存储目录，它根据不同的路径创建不同的文件。这一点和前文的 SharedPreferences、sqflite 相似，都是通过插件实现的。该插件名为 path_provider，下面我们来介绍一下它的用法。

8.3.1　path_provider 简介

首先，我们看一看引入插件的方式，代码如下所示：

```
dependencies:
  path_provider: ^0.5.0+1
```

在 path_provider 里，有 3 种获取目录的方式，我们分别介绍一下。

（1）获取临时目录，方式如下：

```
Directory tempDir = await getTemporaryDirectory();
String tempPath = tempDir.path;
```

（2）获取应用文档目录，方式如下：

```
Directory appDocDir = await getApplicationDocumentsDirectory();
String appDocPath = appDocDir.path;
```

（3）获取外部存储目录，方式如下：

```
Directory externalDir = await getExternalStorageDirectory();
String externalPath = externalDir.path;
```

需要注意的是，在 iOS 平台上没有外部存储目录的概念，这一点在实际开发中需要注意并区分。

8.3.2　一个简单的日记本示例

本例将展示一个简单的日记本的写法。在文本框的输入完成之后，点击保存并生成一个文件，然后马上读取文件的内容并在屏幕上显示。代码如下所示：

```
import 'dart:io';

import 'package:flutter/material.dart';
import 'package:path_provider/path_provider.dart';
```

```dart
class FileProviderPage extends StatefulWidget {
  final InfoStorage storage;
  final String title;

  FileProviderPage({Key key, this.title, this.storage}) : super(key: key);

  @override
  _FileProviderPageState createState() =>
      _FileProviderPageState();
}

class _FileProviderPageState extends State<FileProviderPage> {
  final TextEditingController textTitleEditingController =
      new TextEditingController();
  String _info;

  @override
  void initState() {
    super.initState();
    widget.storage.readInfo().then((String info) {
      setState(() {
        _info = info;
      });
    });
  }

  Future<File> _saveInfo() async {
    setState(() {
      _info = textTitleEditingController.text;
    });

    return widget.storage.writeInfo(_info);
  }

  @override
  Widget build(BuildContext context) {
```

```
    return Scaffold(
      appBar: AppBar(title: Text('一个简单的日记本')),
      body: Column(
        children: <Widget>[
          TextField(
            controller: textTitleEditingController,
            autofocus: true,
            decoration: new InputDecoration(
              hintText: _info,
            ),
          ),
          RaisedButton(
            child: Text('保存'),
            onPressed: _saveInfo,
          ),
          Text(_info ?? ""),
        ],
      ),
    );
  }
}

class InfoStorage {
  Future<String> get _localPath async {
    final directory = await getApplicationDocumentsDirectory();
    return directory.path;
  }

  Future<File> get _localFile async {
    final path = await _localPath;
    return File('$path/info.txt');
  }

  Future<String> readInfo() async {
    try {
      final file = await _localFile;
      String info = await file.readAsString();
      print(info);
```

```
    return info;
  } catch (e) {
    return "error";
  }
}

Future<File> writeInfo(String info) async {
  final file = await _localFile;
  return file.writeAsString(info);
  }
}
```

在上述代码中，我们定义了一个 InfoStorage 类，在该类中可以实现加载文件路径以及文件内容，然后调用类里面的方法来实现相应的日记功能。

完整的代码见 chapter8/flutter_store/lib/filepath_provider.dart。

本章小结

本章提及的几个相关的插件都可以在 Dart Packages 网站上找到，读者可以根据自己的需求并结合实际的项目来查找需要的插件。另外，本章讲解的持久化的 3 种方式，读者可以根据实际的项目情况进行技术选型。

第 9 章 插件与混合工程

目前，纯原生开发的 App 在市场上几乎没有了，使用比较多的是混合技术，比如 Hybrid、ReactNative、Weex。因此，我们将讲解怎样用 Flutter 做出混合型的 App。

Flutter 的工程形态有多种，可以以插件的形式来开发，比如第 8 章所讲到的 shared_preferences、sqflite、path_provider，还可以在原有的 Native 工程中嵌入 Flutter。后者的运用更多一些，这是因为市场上的 App 大多使用 Native 作为壳，然后在里面嵌入 H5 或 Flutter 等技术。由于 Flutter 还在发展过程中，其原生能力还不完善，需要在 Native 的层面以类似插件的形式提供原生服务。本章就让我们来认识并了解一下各种工程的形态。

9.1 package

Flutter 支持开发者开发自己的 package，并发布到 Dart Packages 网站上。这样可以让我们快速构建 App，从而避免了重复造轮子（或者在轮子的基础上造新轮子）的情况。

9.1.1 添加 package 的几种方式

我们先了解一下添加 package 的几种方式。

1. 添加已发布的 package

这种方式在前面几个章节中都接触过。在 Dart Packages 网站上我们搜索相应的 package 名字之后，点击对应的 Installing 选项卡，添加 dependencies 到 pubspec.yaml 即可。例如下面这样：

```
dependencies:
    battery: ^0.3.0+1
```

然后就是安装了，进入工程的主目录，输入命令"flutter packages get"并按一下回车键，只需要稍等片刻，就可以安装成功。

2. 添加未发布的 package

这种情况也非常多见，有时候我们的 package 只是团队内部使用，并不会发布到 Dart Packages 网站上。此时，我们可以在相对应的路径中添加 package，如下所示：

```
dependencies:
plugin1:
        path: ../plugin1/
```

这个 plugin1 目录和我们的项目名是同级的。我们再看一下采用 git 的引入方式，如下所示：

```
dependencies:
  plugin1:
    git:
      url: git://github.com/flutter/plugin1.git
```

上面 git 的引入方式是在根目录中进行的，也可以通过 path 指定相对应的位置，如下所示：

```
dependencies:
```

```
package1:
  git:
    url: git://github.com/flutter/packages.git
    path: packages/package1
```

9.1.2 更新 package

在第一次添加完一个插件并运行"flutter packages get"之后，会生成一个 pubspec.lock 的文件。文件的后缀名是 .lock 文件，而 lock 又代表"锁"的意思，这个文件的大致内容如图 9.1 所示。

```
1   # Generated by pub
2   # See https://www.dartlang.org/tools/pub/glossary#lockfile
3   packages:
4     async:
5       dependency: transitive
6       description:
7         name: async
8         url: "https://pub.flutter-io.cn"
9       source: hosted
10      version: "2.0.8"
11    boolean_selector:
12      dependency: transitive
13      description:
14        name: boolean_selector
15        url: "https://pub.flutter-io.cn"
16      source: hosted
17      version: "1.0.4"
18    charcode:
19      dependency: transitive
20      description:
21        name: charcode
22        url: "https://pub.flutter-io.cn"
23      source: hosted
24      version: "1.1.2"
25    collection:
26      dependency: transitive
27      description:
28        name: collection
29        url: "https://pub.flutter-io.cn"
30      source: hosted
31      version: "1.14.11"
```

图 9.1

在这个文件中记录的都是对应的 package 的描述，里面主要的描述字段是 version，通过"锁定"能确保我们再次执行"flutter packages get"命令时，更新的 package 不会影响原来安装过的 package。如果不"锁定"的话，最新的 package 可能不能兼容我们现有的代码。

有时候，我们就是想更新并覆盖原来的 package，此时我们可以通过"flutter packages upgrade"命令进行升级，这种方式在第 1 章中已经介绍过。

9.1.3 创建自己的 package

当在 Dart Packages 网站上搜不到合适的 package 时，我们可以考虑自己创造一个 package。

首先，可以通过命令行创建一个 package，操作如下：

```
flutter create --org com.example --template=plugin hello
```

在这个命令中，我们增加 org 参数来指定包名，还能通过-i 和-a 分别指定 iOS 和 Android 的开发语言，例如：

```
flutter create --template=plugin -i swift -a kotlin hello
```

当 package 插件创建完成后，将生成以下 3 个文件。

（1）README.md：负责介绍 package。

（2）CHANGELOG.md：介绍 package 每个版本的修改记录。

（3）LICENSE：可以添加 LICENSE 相关的信息。

9.1.4 发布 package

在编写完 package 之后，检查 README.md、CHANGELOG.md、LICENSE 的信息是否有误，如果没有错误，就可以把 package 发布到 Dart Packages 网站上了。下面是具体流程，我们逐一介绍。

首次发布，我们通过命令行先检查一下 package 有没有问题，如下所示：

```
pub publish --dry-run
```

运行完之后，终端系统会给出一些修改意见或存在的问题，如图 9.2 所示。

```
Missing requirements:
* Your pubspec.yaml must have an "author" or "authors" field.
* Your pubspec.yaml is missing a "homepage" field.
Suggestions:
* line 11, column 1 of example/test/widget_test.dart: This package doesn't depe
d on hello_example.
  import 'package:hello_example/main.dart';
  ^^^^^^^^^^^^^^^^^^^^^^^^^^^^^^^^^^^^^^^^
Sorry, your package is missing some requirements and can't be published yet.
For more information, see: https://www.dartlang.org/tools/pub/cmd/pub-lish
```

图9.2

在--dry-run 检查并修改之后，我们可以运行相关的命令并发布，如下所示：

```
flutter packages pub publish
```

当发布完成之后，就可以在 Dart Packages 网站上搜索到发布的 package 了。

9.2 理解Platform Channel

Platform Channel 是 Flutter 端与 Platform 端制定的通信机制，它分为以下 3 种。

（1）BasicMessageChannel：用于传递字符串和半结构化的信息（在大内存数据块传递的情况下使用）。

（2）MethodChannel：用于传递方法的调用（method invocation）。

（3）EventChannel：用于数据流（event streams）的通信。

虽然三种 Platform Channel 用途不同，但都有一个共同点，就是以 BinaryMessager 作为通信工具，下面，我们来具体讲解一下。

9.2.1 消息传递与编解码器

在了解三种 Platform Channel 之前，我们要先对 Flutter 的消息传递和编解码器有一些了解。

Flutter 的消息传递工具是 BinaryMessager，我们通过它与 Platform 建立起通信关系，并且传递的消息格式是二进制的。BinaryMessager 要与三种 Platform Channel 建立通信关系，中间就需要一个"转换过程"，这个过程我们用示意图表示一下，如图 9.3 所示。

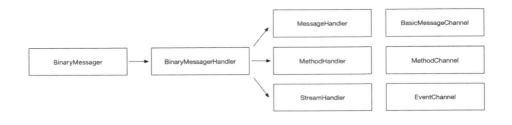

图 9.3

从图 9.3 可以看出，BinaryMessager 要传递下去，还需要经过 BinaryMessagerHandler。BinaryMessagerHandler 是以 Channel name 作为键值生成出来，再被注册到 BinaryMessager 上的。BinaryMessagerHandler 和 BinaryMessager 是一一对应的。二进制格式的消息通过消息编解码器（Codec）解码为能识别的信息，并传递给 Handler 来进行处理。Handler 处理完后，会把结果编码为二进制格式，再通过回调函数返回结果并发送回 Flutter 端。

编码分为两种：MessageCodec 和 MethodCodec。其中，MessageCodec 包括 BinaryCodec、StringCodec、JSONMessageCodec 和 StandardMessageCodec，MethodCodec 包括 JSONMethodCodec 和 StandardMethodCodec。经过消息编解码器处理之后，消息就可以被 Handler（不是 Android 里的 Handler）进行处理了。Handler 有三种，如图 9.3 所示，它与各自的 Channel 有对应关系。Handler 的具体作用如表 9.1 所示。

表 9.1

Handler	作用
MessageHandler	用于处理字符串或者半结构化的消息，对应 BasicMessageChannel
MethodHandler	用于处理方法的调用，对应 MethodChannel
StreamHandler	用于事件流的通信，对应 EventChannel

9.2.2　Platform 数据类型支持

理解了前文的编解码之后，我们看一下 Flutter 默认的消息编解码器，它就是 StandardMessageCodec，目前支持的数据类型如图 9.4 所示。

Dart	Android	iOS
null	null	nil (NSNull when nested)
bool	java.lang.Boolean	NSNumber numberWithBool:
int	java.lang.Integer	NSNumber numberWithInt:
int, if 32 bits not enough	java.lang.Long	NSNumber numberWithLong:
double	java.lang.Double	NSNumber numberWithDouble:
String	java.lang.String	NSString
Uint8List	byte[]	FlutterStandardTypedData typedDataWithBytes:
Int32List	int[]	FlutterStandardTypedData typedDataWithInt32:
Int64List	long[]	FlutterStandardTypedData typedDataWithInt64:
Float64List	double[]	FlutterStandardTypedData typedDataWithFloat64:
List	java.util.ArrayList	NSArray
Map	java.util.HashMap	NSDictionary

图 9.4

图 9.4 详细地解释了 Dart 与 Android、iOS 两者之间的编解码所对应的规则，通过这个对应规则可以让消息进行二进制序列化与反序列化。

9.2.3　MethodChannel 简介

前文我们介绍了消息传递和编码，本节我们了解一下最常用的 MethodChannel。MethodChannel 是 Flutter 与 Platform 之间消息传递的一种，其传递过程是：BinaryMessager→BinaryMessagerHandler→MethodHandler→MethodChannel。从消息传递过程来看，它体现了 Flutter 系统的灵活性。我们看一下消息传递的示意图，如图 9.5 所示。

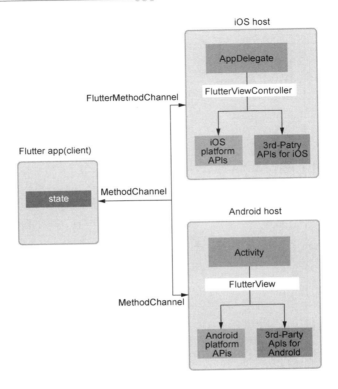

图 9.5

在图 9.5 中，Native 端（iOS 和 Android）为宿主端（host），Flutter 则是客户端（client）。

在 Flutter 中调用 Native 方法时，需要传递的信息是通过平台通道传递到原生端（宿主端）的，Native 收到调用信息后方可执行指定的操作。如有返回的数据，则 Native 会将数据再通过平台通道一并传递给 Flutter，其中消息传递是异步的，这样就确保了消息传递时用户界面不会被阻塞。在客户端，MethodChannel API 可以发送与调用相对应的消息。在图 9.5 中并没有提到 BinaryMessenger，但通过前面小节的学习，我们能理解其中的内部实现原理。

需要注意的是 Platform Channel 并非是安全的线程，所以 MedhodChannel 也一样，它需要确保 Native 宿主端回调函数是在 Platform Thread（Android 和 iOS 的主线程）中被执行的。

9.2.4 SharedPreferences 插件源码解析

在前面的小节中，我们讲解了如何创建一个 package，也介绍了 MethodChannel，本节我们讲解一下 SharedPreferences 插件，从源码入手，理论结合实践进行分析。

回顾一下，在第 8 章中我们实践的例子是一个计数器，在代码中，我们调用了以下方法获取了 SharedPreferences 存储的 counter 值，代码如下所示：

```
SharedPreferences prefs = await SharedPreferences.getInstance();
int counter = (prefs.getInt("counter") ?? 0) + 1;
```

下面我们从 SharedPreferences.getInstance 入手，看一看源码是怎么做的，代码如下所示：

```
static Future<SharedPreferences> getInstance() async {
  if (_instance == null) {
    final Map<Object, Object> fromSystem =
        await _kChannel.invokeMethod('getAll');
    assert(fromSystem != null);
    final Map<String, Object> preferencesMap = <String, Object>{};
    for (String key in fromSystem.keys) {
      assert(key.startsWith(_prefix));
      preferencesMap[key.substring(_prefix.length)] = fromSystem[key];
    }
    _instance = SharedPreferences._(preferencesMap);
  }
  return _instance;
}
```

该方法为静态方法，并通过 getInstance 获取 SharedPreferences 实例。如果在 instance 首次获取的过程中，Flutter 端定义了一个 fromSystem 的 Map 类型，那么将下面 Flutter 定义的 Channel 签名与原生定义的 Channel 进行匹配，然后执行上述代码中的 invokeMethod('getAll')，查找到对应的原生端代码，MethodChannel 的签名在 Flutter 端的定义如下：

```
const MethodChannel _kChannel =
    MethodChannel('plugins.flutter.io/shared_preferences');
```

Android 端也定义了与之匹配的静态常量，代码如下所示：

```
private static final String CHANNEL_NAME = "plugins.flutter.
io/shared_preferences";
```

作为原生的 Android 端（也就是宿主端），定义的插件名称是 SharedPreferencesPlugin，并且实现了 MethodCallHandler 的接口，又复写了 onMethodCall 方法（这个方法用来监听 Flutter 客户端的消息发起），根据 Flutter 源码里的 invokeMethod('getAll')方法，可以查找在 Android 端中调用的方法代码，如下所示：

```
    private Map<String, Object> getAllPrefs() throws IOException {
      Map<String, ?> allPrefs = preferences.getAll();
      Map<String, Object> filteredPrefs = new HashMap<>();
      for (String key : allPrefs.keySet()) {
        if (key.startsWith("flutter.")) {
          Object value = allPrefs.get(key);
          if (value instanceof String) {
            String stringValue = (String) value;
            if (stringValue.startsWith(LIST_IDENTIFIER)) {
              value = decodeList(stringValue.substring
(LIST_IDENTIFIER.length()));
            } else if (stringValue.startsWith(BIG_INTEGER_PREFIX)) {
              String encoded = stringValue.substring(BIG_INTEGER_
PREFIX.length());
              value = new BigInteger(encoded, Character.MAX_RADIX);
            } else if (stringValue.startsWith(DOUBLE_PREFIX)) {
              String doubleStr = stringValue.substring(DOUBLE_
PREFIX.length());
              value = Double.valueOf(doubleStr);
            }
          } else if (value instanceof Set) {
            List<String> listValue = new ArrayList<>((Set) value);
            boolean success =
```

```
            preferences
                .edit()
                .remove(key)
                .putString(key, LIST_IDENTIFIER + encodeList(listValue))
                .commit();
        if (!success) {
          throw new IOException("Could not migrate set to list");
        }
        value = listValue;
      }
      filteredPrefs.put(key, value);
    }
  }
  return filteredPrefs;
}
```

从上面这段代码中，我们知道获取原生 API 里的 android.content.SharedPreferences 类所存放的以"flutter."作为 key 的值，可以通过定义一个 Map 类型的 filteredPrefs 变量，来存放需要存储的 key 和 value 值。再通过 result.success(filteredPrefs)把 Android 端存放的 SharedPreferences 结果作为返回值，返回给 Flutter 客户端，而 Flutter 端会把这个 Map 缓存到 _preferenceCache，然后使用者就可以通过 getInt、getString 等方法从 _preferenceCache 里面取值并返回给应用层了。

请读者思考一个问题,当客户端发起 setInt 时,Android 宿主端和 Flutter 客户端分别又做了什么？读者也可以去看一下其他人发布的库，并分析其源码。

9.3 混合开发

本节我们介绍一下在 Native 工程中如何嵌入 Flutter。这种方式是在面向企业级项目时，大家使用最多的工程创建形式。我们可以在原生应用里面的某一块嵌入 Flutter 界面，反之，也可以在 Flutter 里面嵌入原生的一个界面（View）。

9.3.1 创建 Flutter 模块

在 Native 工程的同级目录下运行以下命令,并创建一个 Flutter 模块,命令如下所示:

```
flutter create -t module flutter_module
```

经过短暂的等待,flutter_module 就会被创建成功,如图 9.6 所示。

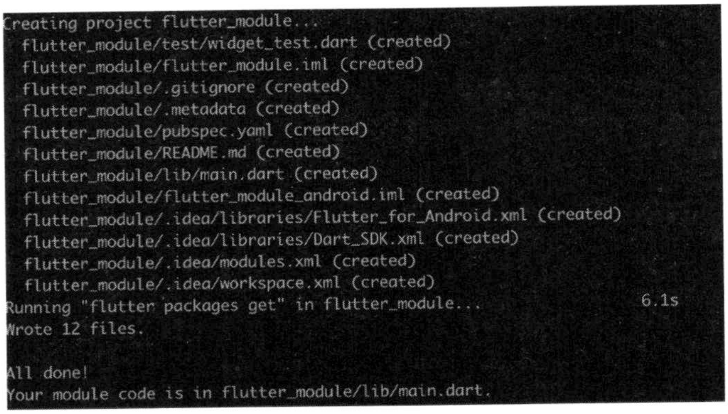

图 9.6

9.3.2 关联原生工程

在需要集成 Flutter 模块的原生工程里执行以下操作步骤。

(1)尽量把原生工程和 Flutter 模块工程放在同一目录下,但这一点不是强制性的。我们最好设定两者在同一目录下,比如 Android 的工程是 AndroidNative,则应与前面创建的 flutter_module 模块工程目录平级。

(2)在 Android 原生工程目录下找到 settings.gradle,并做更改,代码如下所示:

```
include ':app'
setBinding(new Binding([gradle: this]))
0evaluate(new File(
```

```
settingsDir.parentFile,
'flutter_module/.android/include_flutter.groovy'
))
```

（3）需要检查 App 所在的 build.gradle 文件有没有被添加 Java8 编译环境，代码如下所示：

```
compileOptions {
    sourceCompatibility JavaVersion.VERSION_1_8
    targetCompatibility JavaVersion.VERSION_1_8
}
```

（4）在 App 的 build.gradle 文件里引入 flutter 库，代码如下所示：

```
dependencies {
    implementation project(':flutter')
}
```

9.3.3 编写混合工程代码

在关联完原生工程之后，我们需要分别在宿主端和 Flutter 端写一些代码来检验是否成功创建了混合工程。

（1）修改 Android 端的 MainActivity，代码如下所示：

```
public class MainActivity extends AppCompatActivity {
    @Override
    protected void onCreate(Bundle savedInstanceState) {
        super.onCreate(savedInstanceState);
        View flutterView = Flutter.createView(
            MainActivity.this,
            getLifecycle(),
            "route1"
        );
        FrameLayout.LayoutParams layout = new FrameLayout
.LayoutParams (FrameLayout.LayoutParams.MATCH_PARENT,
            FrameLayout.LayoutParams.MATCH_PARENT);
        addContentView(flutterView, layout);
```

```
    }
}
```

上面的 flutterView 可以嵌入任何一个 Android 的 View 里面，读者可以根据实际需要来处理。另外，除了 Flutter.createView 的方式外，还有 FlutterFragment 的方式，代码如下所示：

```
fab.setOnClickListener(new View.OnClickListener() {
  @Override
  public void onClick(View view) {
    FragmentTransaction tx = getSupportFragmentManager().beginTransaction();
    tx.replace(R.id.someContainer, Flutter.createFragment("route1"));
    tx.commit();
  }
});
```

（2）改写 Flutter 的 lib/main.dart 文件，代码如下所示：

```
import 'dart:ui';
import 'package:flutter/material.dart';

void main() => runApp(_widgetForRoute(window.defaultRouteName));

Widget _widgetForRoute(String route) {
  switch (route) {
    case 'route1':
      return MyApp();
    default:
      return Center(
        child: Text('Unknown route: $route', textDirection: TextDirection.ltr),
      );
  }
}
```

需要注意的是，在 swtich/case 逻辑里，route1 需要对应 Android 工程里的 route1 参数，只有对应了之后，Flutter 才会返回对应的 Widget 界面。

在执行完以上两步操作之后，我们在 Android 端运行代码，效果如图 9.7 所示。

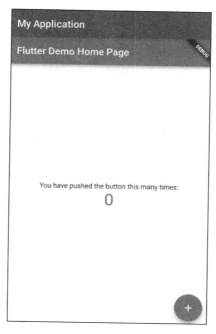

图 9.7

从图 9.7 我们可以看出，Android 嵌入 Flutter 的工程我们已经集成好了。在本例中，Android 嵌入的 Flutter 只用几行代码就实现了，后续要做的事情就是实际项目开发了。我们可以通过 MethodChannel 传递信息，而传递的方式已经在前面的章节中讲解过了，这里不再重复。

9.3.4　热重载混合端代码

在纯 Flutter 工程中，热重载是管用的，但是在混合工程中就失效了。其实，Flutter 官方也想到了这一点，只需要在 Flutter 对应的目录中输入"flutter attach"命令就可以解决，代码如下所示：

```
$ cd some/path/flutter_module
$ flutter attach
Waiting for a connection from Flutter on Nexus 5X...
```

我们在 flutter_module 工程里任意修改一些内容，命令行就会出现对应的信息，如图 9.8 所示。

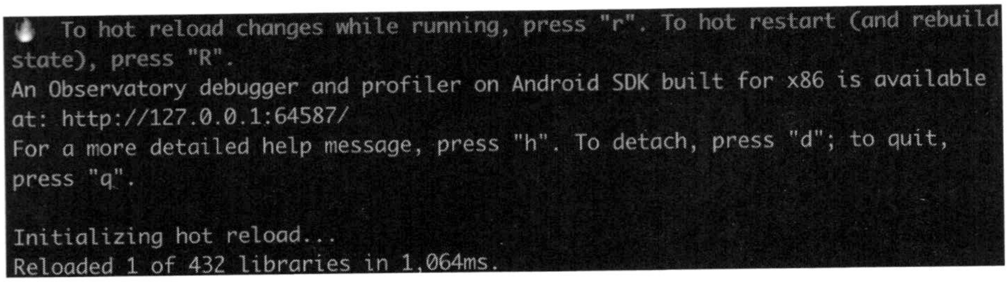

```
To hot reload changes while running, press "r". To hot restart (and rebuild state), press "R".
An Observatory debugger and profiler on Android SDK built for x86 is available at: http://127.0.0.1:64587/
For a more detailed help message, press "h". To detach, press "d"; to quit, press "q".

Initializing hot reload...
Reloaded 1 of 432 libraries in 1,064ms.
```

图 9.8

根据提示内容，如果需要热重载，则按"r"键，如果需要热重启，则按"R"键，按完之后界面就会更新到最新的状态。

9.3.5　aar 模块化打包

在团队开发中，并不是每个人都需要懂 Flutter 开发，很有可能一些人只会 H5 开发，一些人只会 Native 开发。那么，有没有必要让他们都安装 Flutter 开发环境呢？这样其实很麻烦，这时我们可以在 Flutter 中通过 aar 模块化打包的方式来解决这种多人协作的环境问题。

在命令行中，我们进入 flutter_module 的 .android 目录，然后输入以下 gradlew 命令：

```
$ cd .android/
$ ./gradlew flutter:assembleDebug
```

稍等片刻就会出现一个 flutter-debug.aar，该 aar 存放在 ".android/Flutter/build/outputs/aar/" 中。

以上 aar 模块化打包方案是 Flutter 官方给出的。但实际上，很多互联网公司的项目架构更复杂。这时就需要移动端架构师，他们会制定出一套适合公司项目的灵活构建方式，比如闲鱼的 Flutter 技术就有过相关的实践分析。本书最后有一个完整的 fat-aar 打包方式供读者参考,源码在 chapter9/flutter_fat_aar。

9.4 FlutterBoost混合方案

闲鱼 Flutter 团队与谷歌 Flutter 团队有密切的沟通，FlutterBoost 就是闲鱼团队研究出的一套混合方案。那么 FlutterBoost 为我们解决了什么问题呢？我们先来理解一下官方给出的混合方案的基本原理。

9.4.1 框架的由来

Flutter 主要由使用 C++实现的 Flutter Engine 和使用 Dart 实现的 Framework 组成。

其中，Flutter Engine 负责线程管理、Dart VM（虚拟机）状态管理和 Dart 代码加载等工作。

Framework 层是与业务层打交道的，并且通过 Dart 来实现，比如 Widget。一个进程里面只会初始化一个 Dart VM，而一个进程可以有多个 Flutter Engine，多个 Engine 实例共享同一个 Dart VM。

我们看一个业务场景，在 Android 里每初始化一个 FlutterView 就会有一个 Flutter Engine 被初始化，这样就会有新的线程在 Dart 上运行。比如，每个 Activity 绑定一个 FlutterView，那么就会创建多个 Flutter Engine 实例，这样就会出现每个 Flutter Engine 实例加载的代码都独立运行在 isolate 中，这种模式被称为多引擎模式。闲鱼团队总结出了多引擎模式存在以下的问题。

（1）冗余的资源问题。在多引擎模式下每个引擎之间的 isolate 都是相互独立的，但是引擎底层各自维护了图片缓存等比较消耗内存的对象。因此，内存压力将会非常大。

（2）插件注册的问题。插件依赖 Messenger 去传递消息，在多个 FlutterView 情况下，插件的注册和通信将会变得混乱而难以维护，最终会导致消息传递的源头和目标也变得不可控。

（3）Flutter Widget 和 Native 的页面差异化问题。Flutter 的页面是 Widget，Native 的页面是 Activity 或 Fragment。这种差异在进行类似埋点操作时，会增加额外的复杂度。

（4）增加页面之间通信的复杂度，导致多引擎实例也变得更加复杂。

对于此类问题，闲鱼 Flutter 团队与谷歌 Flutter 团队进行了沟通。谷歌给出的建议是，对于连续的 Flutter 页面（Widget）只需要在当前的 FlutterView 中打开，对于间隔的 Flutter 页面可以初始化新的引擎。换言之，闲鱼 Flutter 团队需要优化的是所有绘制窗口并共享同一个 Root isolate。

9.4.2　使用 FlutterBoost 改进

针对上述小节碰到的问题，闲鱼 Flutter 团队经历了不断改进的过程，最后研究出了 FlutterBoost 框架。其设计思想是把 Flutter 容器做成浏览器的样子，然后填写一个页面地址，再由容器去管理页面的绘制。在 Native 端，如果初始化容器，就设置容器对应页面的标志即可。闲鱼 Flutter 团队的核心架构如图 9.9 所示。

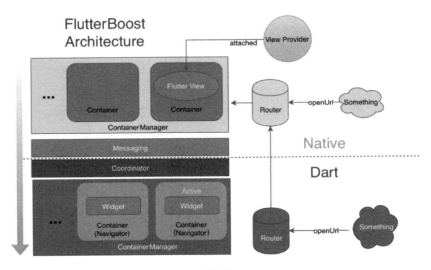

图 9.9

图 9.9 分为 Native 层和 Dart 层。

1. Native 层

Container：包括 Native 容器、平台 Controller、Activity、ViewController。

Container Manager：容器的管理者。

Adaptor：Flutter 的适配层。

Messaging：基于 Channel 的消息通信。

2. Dart 层

Container：Flutter 用来容纳 Widget 的容器，具体实现为 Navigator 的派生类。

Container Manager：Flutter 容器的管理，提供 show、remove 等 API。

Coordinator：包括协调器、接受 Messaging、负责调用 Container Manager 的状态管理。

Messaging：基于 Channel 的消息通信。

通过这样的框架技术整改，闲鱼 Flutter 团队解决了多引擎模式存在的诸多问题。整体就像在使用普通的浏览器访问网页一样，即在 Native 层面进行了收口。笔者亲自测试发现，在 Flutter 最新的 1.5.4 版本中该方案也是可行的，并未发现兼容性的问题。

下面，我们带着疑问来阅读一些 FlutterBoost 相关的源码。

9.4.3 FlutterBoost 源码分析

笔者把 FlutterBoost 的源码中有关 Android 的例子单独运行了一遍，运行效果如图 9.10 所示。

当点击"open flutter fragment page"之后，跳入新的页面，如图 9.11 所示。

图 9.10

图 9.11

我们可以不断地点击，不断地进入 Native 界面，再不断地进入 Flutter 界面，返回就按照添加入栈的顺序依次出栈。这样对于用户来说，是不知道当前界面的实现技术的。应用也避免了在多引擎模式下的各种无用的消耗。我们来分析一下源码是如何实现的。

首先，初始化 FlutterBoostPlugin，再根据 FlutterBoost 接入文档。这里需要初始化插件，代码如下所示：

```
FlutterBoostPlugin.init(new IPlatform() {
    @Override
    public Application getApplication() {
        return MyApplication.this;
    }

    /**
     * 获取应用入口的 Activity,它在应用交互期间应该是一直在栈底的
     * @return
     */
    @Override
```

```java
    public Activity getMainActivity() {
        return MainActivity.sRef.get();
    }

    @Override
    public boolean isDebug() {
        return true;
    }

    @Override
    public boolean startActivity(Context context, String url, int requestCode) {
        return PageRouter.openPageByUrl(context,url,requestCode);
    }

    @Override
    public Map getSettings() {
        return null;
    }
});
```

再看 init 方法的定义，代码如下所示：

```java
public static synchronized void init(IPlatform platform) {
    if (sInstance == null) {
        sInstance = new FlutterBoostPlugin(platform);
        platform.getApplication().registerActivityLifecycleCallbacks(sInstance);
        ServiceLoader.load();
    }
}
```

在初始化的过程中，我们做了以下几件事。

（1）把 IPlatform 的接口实现作为参数传递给了 new FlutterBoostPlugin。

（2）获取应用入口的 Activity，它在应用交互期间一直在栈底。我们通过 registerActivityLifecycleCallbacks 方法绑定 Activity 的生命周期。

（3）注册 NavigationService 服务。

了解了这些之后，启动 MainActivity，有 3 个按钮，分别实现了以下跳转。

（1）"open native page"实现了 Native→Native。

（2）"open flutter page"实现了 Native→Flutter Activity。

（3）"open flutter fragment page"实现了 Native→Flutter Fragment。

其中，第一种跳转"open native page"是 Native 之间的普通跳转。

我们来详细看一下第二种跳转"open flutter page"（第三种跳转和第二种跳转差不多，就不细说了）。

在 FlutterPageActivity 里，调用了 getContainerName 方法，并返回"flutterPage"，然后 Flutter 的 Dart 端会接收到对应的"界面名字"，并对应进行处理，最后通过路由跳转到 flutterPage 的 Widget 界面。在整个路由过程中，Android 端做了以下处理。

（1）FlutterBoost 封装了 BoostFlutterActivity 类。由于前面在 Application 里已经初始化了 FlutterBoostPlugin，因此可以直接在 BoostFlutterActivity 的 onCreate、onPostName、onPause、onDestory 等生命周期方法里，通过 FlutterBoostPlugin 获取 Native 的容器管理类 containerManager，并且对 onContainerCreate、onContainerDisappear、onContainerDestroy 等方法进行监听。以 onContainerCreate 为例，实现代码如下所示：

```
@Override
public PluginRegistry onContainerCreate(IFlutterViewContainer container) {
    assertCallOnMainThread();
    ContainerRecord record = new ContainerRecord(this, container);
    if (mRecords.put(container, record) != null) {
        Debuger.exception("container:" + container.getContainerName()
+ " already exists!");
    return new PluginRegistryImpl(container.getActivity(),
container.getBoostFlutterView());
    }
```

```
        mRefs.add(new ContainerRef(record.uniqueId(),container));
        FlutterMain.ensureInitializationComplete
(container.getActivity().getApplicationContext(), null);
        BoostFlutterView flutterView = FlutterBoostPlugin.viewProvider()
.createFlutterView(container);
        if
(!flutterView.getFlutterNativeView().isApplicationRunning()) {
            String appBundlePath = FlutterMain.findAppBundlePath
(container.getActivity().getApplicationContext());
            if (appBundlePath != null) {
                FlutterRunArguments arguments = new FlutterRunArguments();
                arguments.bundlePath = appBundlePath;
                arguments.entrypoint = "main";
                flutterView.runFromBundle(arguments);
            }
        }
        mInstrument.performCreate(record);
        return new PluginRegistryImpl(container.getActivity(),
container.getBoostFlutterView());
    }
```

在上述 onContainerCreate 代码中，创建了一个 LinkedHashMap，其键（key）为 IFlutterViewContainer，值（value）为 IContainerRecord，存放的是 ContainerRecord 对象。然后，我们通过 viewProvider 创建了 FlutterView，并且通过创建完成的 FlutterView 获取了 appBundlePath。再把 appBundlePath 传递给 FlutterRunArguments，设置 FlutterRunArguments 的 entrypoint 属性为 main，并通过 mInstrument.performCreate 创建一条记录，最后返回 PluginRegistry 的实现类。

同样，onContainerAppear、onContainerDisappear、onContainerDestroy 的实现方式也类似，只不过 mInstrument 调用的方式不同，它们分别调用了 performAppear、performDisappear、performDestroy，实现了 FlutterView 进栈出栈的功能。整个过程是通过 FlutterViewContainerManager 类管理的，而 FlutterViewContainerManager 又是在 FlutterBoostPlugin 类里可以获取的。

在上面这些代码中，最关键的事情是 lutterViewContainerManager 要找到对应 ContainerRecord 的唯一 ID。Flutter Container 可以将其设置为前台可见容器，唯一 ID 的生成机制是：System.currentTimeMillis() + "-" + hashCode()。

（2）BoostFlutterActivity 做的另外一件重要的事就是创建了 FlutterViewStub 的一个类，并且通过 attachFlutterView 关联了 BoostFlutterView。FlutterViewStub 的构造函数如下所示：

```java
public FlutterViewStub(Context context) {
    super(context);

    mStub = new FrameLayout(context);
    mStub.setBackgroundColor(Color.WHITE);
    addView(mStub, new FrameLayout.LayoutParams(
            ViewGroup.LayoutParams.MATCH_PARENT, ViewGroup
.LayoutParams.MATCH_PARENT));

    mSnapshot = new ImageView(context);
    mSnapshot.setScaleType(ImageView.ScaleType.FIT_CENTER);
    mSnapshot.setLayoutParams(new FrameLayout.LayoutParams(
            ViewGroup.LayoutParams.MATCH_PARENT, ViewGroup
.LayoutParams.MATCH_PARENT));

    mCover = createFlutterInitCoverView();
    addView(mCover, new FrameLayout.LayoutParams(
            ViewGroup.LayoutParams.MATCH_PARENT, ViewGroup
.LayoutParams.MATCH_PARENT));

    final BoostFlutterView flutterView = getBoostFlutterView();
    if (!flutterView.firstFrameCalled()) {
        mSplashScreenView = createSplashScreenView();
        addView(mSplashScreenView, new FrameLayout.LayoutParams(
                ViewGroup.LayoutParams.MATCH_PARENT, ViewGroup
.LayoutParams.MATCH_PARENT));
    }
}
```

在构造函数中，我们创建了一个满屏的 FrameLayout 布局，一个屏幕快照（mSnapshot），一个遮罩层（mCover），一个欢迎界面（mSplashScreenView）。

在调用 attachFlutterView 时，传入了 BoostFlutterView 并添加到实例名为 mSub 的 FrameLayout 布局中（如果之前有 BoostFlutterView，则会把之前的先移除掉）。在调用 detachFlutterView 时，则会先调用 flutterView.getBitmap 来获取当前屏幕的 bitmap，并存储到快照中，然后通过 handle 发送消息，再把 BoostFlutterView 移除掉。这里使用快照保存是因为在两个 Flutter 页面之间进行切换的时候，只有一个 FlutterView，所以需要对上一个页面进行截图保存。

（3）经过上面的分析，我们大致清楚了在 Android 端发生一次点击，跳转到 FlutterView 所做的事情了。那么在 Dart 代码端 Flutter 具体做了什么，又是如何响应的呢？我们看一下官方例子的代码，如下所示：

```
FlutterBoost.singleton.registerPageBuilders({ // ... 'flutterPage':
(pageName, params, _) => FlutterRouteWidget(), // ... });
```

在上面的代码中，FlutterBoost 创建了一个单例，还在构造函数中通过 ServiceLoader.load 初始化了 Flutter 端的 Service，并且可以与前面讲的 Android 端的 ContainerRecord 的唯一 ID 匹配。然后通过"FlutterBoost.singleton.registerPageBuilders"传入所需要的路由，再通过 Map 存放各个路由。在 registerPageBuilders 方法里，其实起用了 ContainerCoordinator（协调器），其作用是接收 Messaging 消息，并负责调用 Container Manager 的状态管理，代码如下所示：

```
ContainerCoordinator() {
  NavigationService.listenEvent(onChannelEvent);
}

// 此处省略其他部分代码

void onChannelEvent(dynamic event) {
  if (event is Map) {
    Map map = event as Map;
    final String type = map['type'];

    switch (type) {
      //Handler back key pressed event.
```

```
            case 'backPressedCallback':
              {
                final String id = map['uniqueId'];
                FlutterBoost.containerManager
                    ?.containerStateOf(id)
                    ?.performBackPressed();
              }
              break;
            //Enter foregroud
            case 'foreground':
              {
                FlutterBoost.containerManager?.setForeground();
              }
              break;
            //Enter background
            case 'background':
              {
                FlutterBoost.containerManager?.setBackground();
              }
              break;
            //Schedule a frame.
            case 'scheduleFrame':
              {
                WidgetsBinding.instance.scheduleForcedFrame();
                Future<dynamic>.delayed(Duration(milliseconds: 250),
                    () => WidgetsBinding.instance.scheduleFrame());
              }
              break;
          }
        }
      }
```

根据前面 FlutterBoost 的架构图，我们可以了解源码的大致处理流程。但由于篇幅关系，笔者不能面面俱到，读者可以查看更多相关的源码学习资料。

本章小结

本章介绍了 Flutter 中 package 的构建和发布方式，Platform Channel 的消息传递和编解码，也介绍了 Flutter 原生应用的混搭。在实际项目开发过程中，这些都非常有用。对于 package，我们尽量多阅读一些开源项目的源代码，从中可以学习到很多知识，在 Dart Packages 中就为我们提供了很多人气比较高的 package 插件。还介绍了 FlutterBoost 框架，它弥补了 Flutter 本身的不足，推荐读者在做混合应用时采用该框架。

第 10 章

项目实战

千呼万唤始出来,终于到了项目实战章节了。在本章,笔者把前文内容完整地贯穿起来,介绍在实际项目中 Flutter 的运用。第一个实战例子偏项目(招聘类 App),通过对实战项目的练习大家可以掌握常规的 Flutter 项目是怎样开发的。第二个实战项目是一个异常上报系统,用于对打包后的 App 进行错误日志监控,并把异常上报到服务端。

10.1 实战一:实现一个招聘类App

本节,将使用前文已经学过的知识,构建一个招聘类 App,并将 App 命名为"精英直聘"。

10.1.1 项目需求与技术选型

招聘类 App 需求比较简单,业务模块也比较容易理解。在实战项目中重点实现招聘 App 的几个重要模块,如职位列表、公司列表、公司详情、消息板块、个人中心等。为了让项目环境更加真实,笔者也会自己搭建后端服务。我们先看一下技术选型。

Flutter 端涉及的第三方库，我们将采用以下这些：

- shared_preferences: ^0.4.2
- scoped_model: ^0.3.0
- json_annotation: ^2.0.0
- web_socket_channel: ^1.0.12
- dio: ^2.1.0

服务端会采用 Node.js，使用以下技术：

- Node.js（推荐 8.12.0 版本）
- mongoose 作为数据库（mongoose 库简而言之就是在 Node 环境中操作 MongoDB 数据库的一种便捷的封装，一种对象模型工具，类似 ORM，mongoose 将数据库中的数据转换为 JavaScript 对象以供开发人员在应用中使用）
- Koa2 提供接口服务层
- Vue 作为后端展示层

mongoose 的安装方式可以参考其官方网站。另外，还需要安装 Node 环境，安装方式可以去 Node 官方网站查找（推荐安装 8.12.0 版本）。

这样一来，前端与后端就都有了。在学习 Flutter 的同时，我们把 Node 练习了一遍，这也是大前端时代所需要的技术能力。当安装完 Node.js 和 mongoose 之后，在项目的目录下运行"npm install"安装所需要的依赖库，然后运行"npm run dev"就可以启动服务端项目了。

10.1.2 服务端设计

在服务端设计中，数据库是很重要的一块，它负责持久化。在客户端发起请求的时候，会读取数据库并把结果通过服务端的 API 返回给客户端，由于 API 接口是从 mongoose 数据库里面读取的，因此，我们看一下需要哪几类 mongoose 数据库关系表。

1. company（公司）

```
const companySchema = new Schema({
  id: {
    unique: true,
    type: String
  },
  company: String,
  logo: String,
  info: String,
  hot: String,
  // ...
})
```

2. companyDetail（公司详情）

```
const companyDetailSchema = new Schema({
  id: {
    unique: true,
    type: String
  },
  inc: String,
  companyImgsResult: [String],
  tagsResult: [String],
  // ...
})
```

3. job（职位）

```
const jobSchema = new Schema({
  id: {
    unique: true,
    type: String
  },


  title: String,
  salary: String,
```

```
  company: String,
  info: String,
  category: String,
  head: String,
  publish: String,
  link: String,
  // ...
})
```

4. message（消息）

```
const messageSchema = new Schema({
  id: {
    unique: true,
    type: String
  },
  name: String,
  message: String,
  head: String,
  time: String,
  // ...
})
```

在 Node.js 服务启动时，我们通过以下代码来初始化并创建数据库，如下所示：

```
exports.initSchemas = () => {
  glob.sync(resolve(__dirname, './schema', '**/*.js')).forEach(require)
}
```

接下来是初始化 mongoose 数据库连接，这一步至关重要，先看一下代码是如何实现的：

```
exports.connect = () => {
  let maxConnectTimes = 0

  return new Promise((resolve, reject) => {
    if (process.env.NODE_ENV !== 'production') {
```

```
  mongoose.set('debug', true)
}

mongoose.connect(db)

mongoose.connection.on('disconnected', () => {
  maxConnectTimes++

  if (maxConnectTimes < 5) {
    mongoose.connect(db)
  } else {
    throw new Error('数据库挂了吧, 快去修吧少年')
  }
})

mongoose.connection.on('error', err => {
  console.log(err)
  maxConnectTimes++

  if (maxConnectTimes < 5) {
    mongoose.connect(db)
  } else {
    throw new Error('数据库挂了吧, 快去修吧少年')
  }
})

mongoose.connection.once('open', () => {
  resolve()
  console.log('MongoDB Connected successfully!')
})
})
}
```

如代码所示，我们建立并返回了一个 Promise 对象，并且通过 mongoose.connect 尝试连接数据库。在数据库因为某种原因"挂掉"的时候（disconnected）或者出错（error）的时候，可以尝试重连数据库，前提是正常连上 mongoose（即 open 状态）。

在了解了数据库建模之后，我们看一下服务端提供的对 API 接口的封装，即客户端 Flutter 通过访问 API 服务端接口返回数据。这里，我们提供了 Koa2 框架（类似 Java 里的 SpringMVC 等框架），想学习 Node 中后台开发的读者可以深入了解一下这款框架。

这里我们说一下基本原理：在 Flutter 端访问服务端时，Koa2 会通过当前 http 请求的上下文（context）获取请求参数，并创建 mongoose 数据库实例，通过查询数据库之后，把查询结果通过当前的 context 返回至 response 的 JSON 响应数据结构之中。我们以获取职位列表为例，看一下代码，如下所示：

```
// 职位列表
router.get('/jobs/list/:page', async (ctx, next) => {
  const curPage = Number(ctx.params.page)
  const pageSize = 10
  const Job = mongoose.model('Job')
  const total = await Job.find({}).count()
  const totalPage = Math.floor((total + pageSize - 1) / pageSize)
  const hasNextPage = curPage < totalPage ? true : false
  const jobs = await Job.find({}).sort({
    'meta.createdAt': -1
  }).skip((curPage - 1) * pageSize).limit(pageSize)

  ctx.body = {
    'data': {
      jobs,
      'pages': {
        curPage,
        totalPage,
        hasNextPage,
        total
      }
    }
  }
})
```

如上述代码所示，Koa2 路由访问的地址参数是 "'/jobs/list/:page'"，实际上用户访问的是 "https://192.168.11.104:3000/jobs/list/1"。其中 URL 的前缀需要改成自己部署的 IP 地址，可以在 server/index.js 文件里找到 host，对

相应的值进行修改。":page"则是页面的动态参数，表示当前查询的页数，最后把查询结果通过 ctx.body 作为响应返回，这样就可以通过 Flutter 对响应结果进行解析了。其他的 API 接口也类似，就不一一细说了。

10.1.3　Flutter 基础架构

根据项目的业务复杂度，笔者把项目进行了目录层级的划分，如图 10.1 所示。

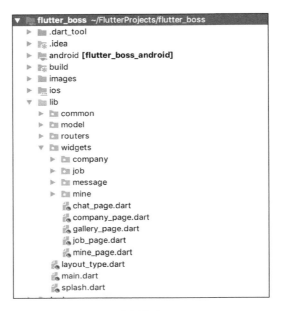

图 10.1

本例暂时不需要用到与 Native 相关的插件代码，因此只需要关心 lib 目录下的 Flutter 文件代码即可。每一级对应的目录含义如下所示。

（1）common：公共部分，存放公共配置、工具类、公共组件。

（2）model：实体类，用 json_serializable 工具生成。

（3）routers：路由相关的存放。

（4）widgets：存放项目里的一级页面和二级页面。

（5）最外层存放 main.dart 和 splash.dart，表示启动画面和主界面。

如何设计项目的存放目录，是没有绝对标准的，通常取决于业务与技术的选型。再者，随着业务的增长以及 Flutter 技术的不断完善，项目也是不断演变的。笔者在接到一个新项目之后，通常会整理出各个业务的模块，并且在有必要的情况下，会画出思维导图来帮助自己理解项目。我们来看一下这个项目中几个重要的模块是如何完成的。

10.1.4　启动页面

启动画面基本上是每一个 App 都具备的，它在 App 启动时显示，会初始化一些数据。其中一个重要作用是获取一些基础数据，比如最新的版本接口。我们来看一下本案例的启动画面，如图 10.2 所示。

图 10.2

启动画面还是比较简洁的，代码如下所示：

```
class SplashPage extends StatefulWidget {
  @override
  SplashState createState() => new SplashState();
}

class SplashState extends State<SplashPage> {
  Timer _t;

  @override
  void initState() {
    super.initState();
    _t = new Timer(const Duration(milliseconds: 1500), () {
      try {
        Navigator.of(context).pushAndRemoveUntil(
          PageRouteBuilder<Null>(
            pageBuilder: (BuildContext context, Animation<double> animation,
                Animation<double> secondaryAnimation) {
              return AnimatedBuilder(
                animation: animation,
                builder: (BuildContext context, Widget child) {
                  return Opacity(
                    opacity: animation.value,
                    child: new MainPage(title: '精英直聘'),
                  );
                },
              );
            },
            transitionDuration: Duration(milliseconds: 300),
          ),
          (Route route) => route == null);
      } catch (e) {}
    });
  }

  @override
  void dispose() {
```

```
    _t.cancel();
    super.dispose();
  }

  @override
  Widget build(BuildContext context) {
    return new Material(
      color: Config.GLOBAL_COLOR,
      child: Container(
        alignment: Alignment(0, -0.3),
        child: new Text(
          "精英直聘",
          style: new TextStyle(
              color: Colors.white, fontSize: 50.0, fontWeight:
FontWeight.bold),
        ),
      ),
    );
  }
}
```

上述代码完成了对启动画面的设置，然后通过 Timer 创建了一个延时等待，在等待 1500ms 之后通过路由的 pushAndRemoveUntil 方法进入 App 首页。这个方法的使用已经在路由相关章节介绍过了，它会在进入首页之后把启动页从路由栈里面移除掉，并且在路由切换过程中，使用动画效果（本例用了一个简单的 Opacity 渐入动画效果）。

在实际项目中，如果先使用的 Native 项目集成 Flutter，那么还是建议把启动画面放在 Native 端。

10.1.5　使用 dio 实现网络请求

在网络相关的章节中我们介绍了 dio 的使用方法，它是一个功能非常强大的网络框架，不仅支持简单的 get/post 请求，还支持拦截器、文件上传和分段下载等。下面我们介绍一下在实际项目中比较有用的做法，即对网络请求的封装。在项目中，我们用了单例模式，代码如下所示：

```
class HttpUtil {
  static final HttpUtil _instance = HttpUtil._internal();
  Dio _client;

  factory HttpUtil() => _instance;

  HttpUtil._internal() {
    if (null == _client) {
      BaseOptions options = new BaseOptions(
        baseUrl: "${Config.BASE_URL}",
        connectTimeout: 1000 * 10,
        receiveTimeout: 3000,
      );
      _client = new Dio(options);
    }
  }

  Future<Response<Map<String, dynamic>>> get(String path,
[Map<String, dynamic> params]) async {
    Response<Map<String, dynamic>> response;
    if (null != params) {
      response = await _client.get(path, queryParameters: params);
    } else {
      response = await _client.get(path);
    }
    return response;
  }
}
```

这样，我们就可以通过一个请求的工具类访问服务端的 API 了。我们封装了 get 方式的请求，它分为带参数和不带参数两种，如果有需要，那么也可以封装 post 方法。我们指定了 baseUrl 和 connectTimeout 的参数设置，这样更便于全局的使用。有了这些对于基础配置的修改，就能在开发效率上有明显提升，比如，对于接口地址的访问，只需要在配置完的 bash 中加上 URL 后缀即可。这样做也便于在测试环境和生产环境中切换 URL 前缀。

10.1.6　公司列表与详情实现

项目中有职位列表和公司列表。我们以公司列表为例说明一下如何以组件的方式编写这个模块，代码如下所示：

```
body: new Center(
  child: FutureBuilder(
    future: _fetchCompanyList(),
    builder: (context, AsyncSnapshot snapshot) {
      switch (snapshot.connectionState) {
        case ConnectionState.none:
        case ConnectionState.waiting:
          return new CircularProgressIndicator();
        default:
          if (snapshot.hasError)
            return new Text('Error: ${snapshot.error}');
          else
            return _createListView(context, snapshot);
      }
    },
  ),
),
```

以上代码用到了 FutureBuilder 组件。在 future 参数中，我们传入了获取列表数据的方法，方法名是"_fetchCompanyList()"，抓取的数据就是服务端中 Koa2 的 API 接口所提供的数据。"_fetchCompanyList()"方法的代码如下所示：

```
Future<List<Company>> _fetchCompanyList() async {
  List<Company> companyList = List<Company>();
  Response<Map<String, dynamic>> response =
      await HttpUtil().get('/company/list/1');
  if (response.statusCode == 200) {
    Map<String, dynamic> result = response.data;
    for (dynamic data in result['data']['companies']) {
      Company companyData = Company.fromJson(data);
```

· 275 ·

```
          companyList.add(companyData);
      }
   }
   return companyList;
}
```

在这段方法中，我们用到了 dio 的网络库的封装类，它只需要简单地调用 "await HttpUtil().get('/company/list/1');" 就可以返回结果，再通过 JSON 解析，把对应的实体类放到 List 集合里面，就可以为列表展示起到作用。我们再回到 FutureBuilder 来看看，如果网络请求正常，则会调用 "_createListView" 方法，该方法通过 ListView.builder 创建了列表，用来展示公司列表数据，代码如下所示：

```
      Widget _createListView(BuildContext context, AsyncSnapshot
snapshot) {
      List<Company> companyList = snapshot.data;
      return ListView.builder(
        key: new PageStorageKey('company-list'),
        itemCount: companyList.length,
        itemBuilder: (BuildContext context, int index) {
          return CompanyItem(
            onPressed: () {
              Navigator.push(
                context,
                MaterialPageRoute(
                  // fullscreenDialog: true,
                  builder: (context) => CompanyDetailPage(
                      company: companyList[index], heroLogo:
"heroLogo${index}"),
                ),
              );
            },
            company: companyList[index],
            heroLogo: "heroLogo${index}",
          );
        },
      );
    }
```

通过 itemBuilder，我们返回了 CompanyItem，这就是封装的一个 CompanyItem 组件，它是对单条数据的描述，代码如下所示：

```
class CompanyItem extends StatelessWidget {
  final Company company;
  final String heroLogo;

  CompanyItem({Key key, this.company, this.onPressed, @required this.heroLogo})
      : super(key: key);
  VoidCallback onPressed;

  @override
  Widget build(BuildContext context) {
    return new GestureDetector(
      onTap: onPressed,
      child: new Container(
        margin: const EdgeInsets.only(bottom: 10.0),
        padding: const EdgeInsets.only(
            left: 18.0, top: 10.0, right: 18.0, bottom: 10.0),
        color: Colors.white,
        child: new Column(
          crossAxisAlignment: CrossAxisAlignment.start,
          mainAxisSize: MainAxisSize.min,
          children: <Widget>[
            new Row(
              children: <Widget>[
                Padding(
                  padding: EdgeInsets.only(right: 20.0),
                  child: Hero(
                    tag: heroLogo,
                    child: Image.network(
                      company.logo,
                      width: 40,
                    ),
                  ),
                ),
                new Column(
```

```
                    crossAxisAlignment: CrossAxisAlignment.start,
                    children: <Widget>[
                      Padding(
                        padding: EdgeInsets.only(bottom: 6.0),
                        child: Text(
                          company.company,
                          style: new TextStyle(color: Colors.black, fontSize: 16),
                        ),
                      ),
                      Text(
                        company.info,
                        style: new TextStyle(color: Colors.grey, fontSize: 12),
                      ),
                    ],
                  ),
                ],
              ),
              new Container(
                decoration: new BoxDecoration(
                    color: new Color(0xFFF6F6F8),
                    borderRadius: new BorderRadius.all(new Radius.circular(6.0))),
                padding: const EdgeInsets.only(
                    top: 3.0, bottom: 3.0, left: 8.0, right: 8.0),
                margin: const EdgeInsets.only(top: 12.0),
                child: Text(
                  company.hot,
                  style: new TextStyle(color: new Color(0xFF9fa3b0)),
                ),
              ),
            ],
          ),
        ),
      );
  }
}
```

在上面的代码中，运用了 Column、Padding、Image 等组件，灵活地构建出了公司的 item 数据，效果如图 10.3 所示。

图 10.3

就这样，一个列表完成了。通过点击列表具体内容，可以到达公司的详情页面，这个动作用到了动态路由和 Hero 效果（需要列表和详情页设置的 tag 一致）。在详情页中，有几个重点的技术细节需要说一下，其整体效果如图 10.4 所示。

图 10.4

这里，在列表中传入了 logo，并且是一个比较大的背景图，其实现方式如下所示：

```
body: Container(
  decoration: BoxDecoration(
    image: DecorationImage(
      colorFilter: new ColorFilter.mode(
          Colors.black.withOpacity(0.1), BlendMode.dstATop),
      fit: BoxFit.cover,
      image: new NetworkImage(widget.company.logo),
      alignment: Alignment.center),
    ),
    child: _companyDetailView(context),
  ),
```

这个效果主要是运用 BoxDecoration 中的 DecorationImage 来实现的，fit 方式被设置为 BoxFit.cover，对齐方式为"居中"即可达到这样的效果。

再看一个效果的实现方式。在滑动带有公司详情介绍的滚动区域时，如果是往上滑动，则会把标题顶上去，如图 10.5 所示。

图 10.5

我们看一下该效果是如何实现的，代码如下所示：

```
_scrollListener() {
  setState(() {
    if (_scrollController.offset < 56 && _isShow) {
      _isShow = false;
    } else if (_scrollController.offset >= 56 && _isShow == false) {
      _isShow = true;
    }
  });
}

@override
void initState() {
  _scrollController = ScrollController();
  _scrollController.addListener(_scrollListener);
  super.initState();
}
```

在这段代码中，我们为 CustomScrollView 加上了"_scrollController"的监听，通过滚动区域的 offset 值来判断顶部的标题是否被隐藏。需要注意的是，我们要在 dispose 中对"_scrollController"进行"销毁"，代码如下所示：

```
@override
void dispose() {
  _scrollController.removeListener(_scrollListener);
  _scrollController.dispose();
  super.dispose();
}
```

其他的组件效果都是比较容易实现的，读者可以查看源代码进行了解。

10.1.7　用 WebSocket 实现聊天模块

在项目中，另一个需要掌握的知识点是 WebSocket 的实现。我们先补充一下 WebSocket 的相关知识。

WebSocket 可以让客户端和服务器之间的数据交换变得更加简单，它允许服务器主动向客户端推送数据。在 WebSocket 的 API 中，客户端和服务器只需要完成一次"握手"，两者之间就可以直接创建持久性的连接，并进行双向数据传输。这样，客户端和服务器之间就形成了一条快速通道，两者之间就可以直接互相传送数据了。

这里，我们用到了 web_socket_channel，在安装该包之后就可以使用它了。我们点击聊天模块，就可以从聊天列表进入聊天历史记录模块，当编辑完消息并点击发送之后，就可以把消息发送出去。当然，WebSocket 在服务端的实现还是需要一些成本的，所以，此处用了一个免费的测试版 WebSocket。我们可以访问 WebSocket 官方旗下相关网址，尝试在浏览器的网页里发送一条消息。结果如图 10.6 所示。

图 10.6

从图 10.6 中右侧区域可以看到，我们发送出去的数据，服务端原封不动地返了回来。接下来，我们将模拟这个过程并在 Flutter 上实现它，最终效果如图 10.7 所示。

要实现上面的展示效果，有以下几个技术点需要实现。

1. 初始化 WebSocketChannel

代码如下所示：

```
WebSocketChannel channel = IOWebSocketChannel.connect
('ws://echo.websocket.org');
```

图 10.7

2. 在 initState 初始化方法里监听 channel 传递来的消息

代码如下所示：

```
channel.stream.listen((msgString){
  Message msg = Message.fromJson(jsonDecode(msgString));
  msg.type = 'receive'; setState(() => _msgList.add(msg));
});
```

这样监听之后，我们就可以往消息记录里添加数据了。

3. 绘制消息列表和发送数据的界面

这里用到了接收消息和发送消息两个组件，它们分别是 message_receive_item 和 message_send_item，二者的代码比较长，此处就看一下接收消息的组件，代码如下所示：

```
class MessageReceiveItem extends StatelessWidget {
  final Message message;

  MessageReceiveItem({Key key, this.message, this.onPressed}) : super(key: key);
  VoidCallback onPressed;

  @override
  Widget build(BuildContext context) {
    return new GestureDetector(
      onTap: onPressed,
      child: new Container(
        margin: const EdgeInsets.only(bottom: 10.0),
        padding: const EdgeInsets.only(
            left: 18.0, top: 10.0, right: 18.0, bottom: 10.0),
        color: Colors.white,
        child: Row(
          crossAxisAlignment: CrossAxisAlignment.start,
          mainAxisSize: MainAxisSize.min,
          children: <Widget>[
            CircleAvatar(
              backgroundImage: NetworkImage(message.head),
              radius: 25,
            ),
            Column(
              crossAxisAlignment: CrossAxisAlignment.start,
              children: <Widget>[
                new Padding(
                  padding: EdgeInsets.only(
                    left: 15,
                  ),
                  child: LimitedBox(
                    maxWidth: MediaQuery.of(context).size.width * 0.6,
                    child: Container(
                      padding: EdgeInsets.all(10.0),
                      decoration: BoxDecoration(
                        borderRadius: BorderRadius.circular(10.0),
```

```
                    color: Color.fromARGB(100, 211, 211, 211),
                  ),
                  child: Text(
                    message.message,
                    style: new TextStyle(color: Colors.black,
fontSize: 16),
                  ),
                ),
              ),
            ),
          ],
        ),
      ],
    ),
  ),
);
  }
}
```

这里用到了 LimitedBox 的组件，设置了聊天内容显示的最大宽度是 60%。聊天底部的发送按钮需要考虑对 iOS 的兼容性，在 iPhoneX 手机上，它需要被放在 SafeArea 组件里。

4. 通过 channel.sink.add 发送数据

代码如下所示：

```
void _sendMessage() {
if (_controller.text.isNotEmpty) {
  final Message msg = Message()
    ..name = widget.message.name
    ..head = widget.message.head
    ..message = _controller.text
    ..type = 'send';
  setState(() => _msgList.add(msg));
  channel.sink.add(jsonEncode(msg.toJson()));
}
}
```

我们使用上面的代码做了一个判断，也就是消息内容如果不为空，则创建一条消息，并添加到聊天列表里，这表示是自己发送出去的。然后，通过"channel.sink.add(jsonEncode(msg.toJson()))"把自己的消息发送到"ws://echo.websocket.org"，根据前面达成的目标，在发出这条消息后它就会原封不动地被返回。

5. 不用时的销毁

代码如下所示：

```
channel@override
void dispose() {
  channel.sink.close();
  super.dispose();
}
```

10.2 实战二：实现异常上报系统

目前市场上大部分商业的 App 都有错误日志系统和埋点系统。这些系统功能都非常实用，比如错误日志系统就是在 App 端收集相关日志并且上报到服务端的。对异常进行上报的目的是为了更好地收集 App 的一些线上问题，提早发现并解决这类问题，避免造成损失。

那么，在 Flutter 中能否实现错误日志系统的搭建呢？答案是肯定的。Flutter 官方列举了一个错误日志的 Demo，但上报的是国外的服务器。下面我们来介绍一下错误日志系统在 Flutter 中的实现原理，并且完整地实现一个简易的错误日志服务器。

10.2.1 实现原理

虽然我们通过简单的方法来实现错误日志系统，但实际情况会复杂很多。整个错误日志系统实现的大致思路如下所示。

（1）创建一个 Flutter 端异常捕获的插件，并写入捕获代码。

（2）把 Plugin 集成到现有的 Flutter 应用里，让 Flutter 应用支持 Plugin 捕获日志。

（3）通过 MethodChannel 注册 Platform 的 Channel 名称，以便在 Flutter 发生异常时可以传递给 Native 端。

（4）在 Native 端收集这些日志之后，再传递给服务器端。

10.2.2　FlutterError.onError 和 Zone

Flutter 官方给我们介绍了 FlutterError 的静态方法，这个方法是 onError。在查看源码后，可以知道其默认处理方式如下所示：

```
static FlutterExceptionHandler onError = dumpErrorToConsole;
```

通过这段代码我们了解到，在 Flutter 中，错误信息直接被放在控制台输出了。这也就是我们平时在开发过程中碰到错误信息时，能在控制台打印出来的原因。

onError 这个方法是可以被重写的，官方的例子如下所示：

```
FlutterError.onError = (FlutterErrorDetails details) {
  if (isInDebugMode) {
    FlutterError.dumpErrorToConsole(details);
  } else {
    Zone.current.handleUncaughtError(details.exception, details.stack);
  }
};
```

上面的代码实现很简单，它用 isInDebugMode 判断了是否是 Debug 环境。如果是，则用 FlutterError.dumpErrorToConsole 方法在控制台输出。

在 Dart 语言中，异常分为两类：即同步异常和异步异常。同步异常可以通过 try catch 来捕获，而带有 Future 这种异步的语句所发生的异常是不能通过 try catch 来获取到的，这时候就需要用到 Zone。

在 Dart 语言里有一个 runZoned 方法，负责为对象指定 Zone，它可以捕获所有未处理的异常。

代码如下所示：

```
runZoned<Future<Null>>(() async {
  runApp(MaterialApp(home: TextViewExample()));
}, onError: (error, stackTrace) async {
  await FlutterCrashReport()
      .reportCrash(error, stackTrace);
});
```

该方法在 runApp 外面包了一层 runZoned，并通过 onError 对异常进行捕获处理。这时，就可以在 onError 里处理错误异常上报的行为。

10.2.3 异常上报 Flutter 的实现

下面，我们就根据前面讲解的内容，来实现实战项目中 Flutter 端日志的截获。首先通过命令创建出一个插件，代码如下所示：

```
class FlutterCrashReport {
  static const MethodChannel _channel =
      const MethodChannel('flutter_crash_report');

  AndroidDeviceInfo _cachedAndroidDeviceInfo;

  static final FlutterCrashReport _singleton =
FlutterCrashReport._internal();

  factory FlutterCrashReport() => _singleton;

  FlutterCrashReport._internal();

  // 初始化错误日志上报系统
  Future<AndroidDeviceInfo> get androidInfo async =>
      _cachedAndroidDeviceInfo ??=
          AndroidDeviceInfo._fromMap(await
```

```
_channel.invokeMethod('deviceInfo'));

  Future<void> onError(FlutterErrorDetails details) async {
    AndroidDeviceInfo androidInfo = await this.androidInfo;
    print('Running on ${androidInfo.model}');

    final data = {
      'device': androidInfo.model,
      'message': details.exception.toString(),
      'cause': details.stack == null ? 'unknown' :
_cause(details.stack),
      'trace': details.stack == null ? [] :
_traces(details.stack),
    };
    return await _channel.invokeMethod('reportCrash', data);
  }

  Future<void> reportCrash(dynamic error, StackTrace stackTrace)
async {
    AndroidDeviceInfo androidInfo = await this.androidInfo;
    print('Running on ${androidInfo.model}');

    final data = {
      'device': androidInfo.model,
      'message': error.toString(),
      'cause': stackTrace == null ? 'unknown' : _cause(stackTrace),
      'trace': stackTrace == null ? [] : _traces(stackTrace),
    };

    return await _channel.invokeMethod('reportCrash', data);
  }

  Future<void> logException(dynamic exception, StackTrace
stackTrace) {
    return reportCrash(exception, stackTrace);
  }

  Future<void> log(String msg, {int priority, String tag}) async {
```

```
      if (priority == null && tag == null) {
        await _channel.invokeMethod('log', msg);
      } else {
        await _channel.invokeMethod('log', [priority, tag, msg]);
      }
    }

    List<Map<String, dynamic>> _traces(StackTrace stack) =>
        Trace.from(stack).frames.map(_toTrace).toList(growable: false);

    String _cause(StackTrace stack) =>
Trace.from(stack).frames.first.toString();

    Map<String, dynamic> _toTrace(Frame frame) {
      final List<String> tokens = frame.member.split('.');

      return {
        'library': frame.library ?? 'unknown',
        'line': frame.line ?? 0,
        'method': tokens.length == 1 ? tokens[0] :
tokens.sublist(1).join('.'),
        'class': tokens.length == 1 ? null : tokens[0],
      };
    }
  }
```

在上述代码中，我们创建了 MethodChannel。在客户端中，我们收集了 device、message、cause、trace 等与错误相关的信息，这些信息作为 Map 对象被传入 Native 端。当然，读者可以根据实际的项目情况添加更多的收集字段。在实现 Flutter Client 之后，我们来看一下 Android 端是如何接收信息的。

10.2.4　异常上报 Android 端的实现

在 Android 端，我们实现了两个 methodCall 传递过来的参数，一个是

deviceInfo,另一个是 reportCrash。其中,deviceInfo 是在 Flutter Client 端发起请求的。我们通过 Android 端获取 devcieInfo 设备相关的信息。reportCrash 则是在客户端出现异常的情况下,传递给服务端数据的。部分代码如下所示:

```java
    @Override
    public void onMethodCall(MethodCall methodCall, Result result) {
        if (methodCall.method.equals("deviceInfo")) {
            Map<String, Object> build = new HashMap<>();
            build.put("board", Build.BOARD);
            build.put("bootloader", Build.BOOTLOADER);
            build.put("brand", Build.BRAND);
            build.put("device", Build.DEVICE);
            build.put("display", Build.DISPLAY);
            build.put("fingerprint", Build.FINGERPRINT);
            build.put("hardware", Build.HARDWARE);
            build.put("host", Build.HOST);
            build.put("id", Build.ID);
            build.put("manufacturer", Build.MANUFACTURER);
            build.put("model", Build.MODEL);
            build.put("product", Build.PRODUCT);
            if (Build.VERSION.SDK_INT >= Build.VERSION_CODES.LOLLIPOP) {
                build.put("supported32BitAbis", Arrays.asList(Build.SUPPORTED_32_BIT_ABIS));
                build.put("supported64BitAbis", Arrays.asList(Build.SUPPORTED_64_BIT_ABIS));
                build.put("supportedAbis", Arrays.asList(Build.SUPPORTED_ABIS));
            } else {
                build.put("supported32BitAbis", Arrays.asList(EMPTY_STRING_LIST));
                build.put("supported64BitAbis", Arrays.asList(EMPTY_STRING_LIST));
                build.put("supportedAbis", Arrays.asList(EMPTY_STRING_LIST));
            }
            build.put("tags", Build.TAGS);
            build.put("type", Build.TYPE);
```

```java
                build.put("isPhysicalDevice", !isEmulator());
                build.put("androidId", getAndroidId());

                Map<String, Object> version = new HashMap<>();
                if (Build.VERSION.SDK_INT >= Build.VERSION_CODES.M) {
                    version.put("baseOS", Build.VERSION.BASE_OS);
                    version.put("previewSdkInt", Build.VERSION.PREVIEW_SDK_INT);
                    version.put("securityPatch", Build.VERSION.SECURITY_PATCH);
                }
                version.put("codename", Build.VERSION.CODENAME);
                version.put("incremental", Build.VERSION.INCREMENTAL);
                version.put("release", Build.VERSION.RELEASE);
                version.put("sdkInt", Build.VERSION.SDK_INT);
                build.put("version", version);

                result.success(build);
            } else {
                onInitialisedMethodCall(methodCall, result);
            }
        }

    void onInitialisedMethodCall(MethodCall call, MethodChannel.Result result) {
            switch (call.method) {
                case "reportCrash":
                    final Map<String, Object> exception = (Map<String, Object>) call.arguments;
                    JSONObject obj = new JSONObject(exception);
                    reportToServer(obj.toString());
                    result.success(null);
                    break;
                default:
                    result.notImplemented();
                    break;
            }
    }
```

在获取了 reportCrash 内容之后，再调用 reportToServer 方法来实现给服务端传递日志数据的步骤。这里用到了 Android 里面的 Okhttp 框架，它把相关的错误信息以 post 方式发送到服务端。代码片段如下所示：

```java
private void reportToServer(String json) {
    RequestBody body = RequestBody.create(JSON, json);
    Log.e("捕获到Flutter异常: ", json);
    Request request = new Request.Builder()
            .url("http://172.20.10.3:3000/log")
            .post(body)
            .build();
    CommonOkHttpClient.getOkHttpClient().newCall(request).enqueue(new Callback() {
        @Override
        public void onFailure(Call call, IOException e) {
            Log.e("response onFailure", call.request().body().toString());
        }

        @Override
        public void onResponse(Call call, Response response) throws IOException {
            Log.e("response", response.body().string());
        }
    });
}
```

需要注意的是这里的 URL 需要修改成实际的 IP 地址，并且与下面介绍的服务端启动后的 IP 地址一致。

10.2.5 服务端接收异常上报

在接收到 Android 端的 post 请求参数后，服务端会把相关的错误日志存入数据库，便于开发者查看和修复。

在这里先介绍一下服务端使用的技术。

（1）主体上采用 Node.js 实现。

（2）Koa2 作为中间层来接收 Android 端发来的 post 请求。

（3）用 mongoose 的数据库进行了日志存储。

读者也可以直接跳过服务端的内容，直接在对应的项目"flutter-log-server"环境下执行"npm install"，然后通过命令"npm run dev"运行。

我们看一下在 Node 端项目启动时的重要配置，代码如下所示：

```
const Koa = require('koa')
const consola = require('consola')
const { Nuxt, Builder } = require('nuxt')
const { connect, initSchemas } = require('./database/init')
const router = require('./routes')

const app = new Koa()
// const host = process.env.HOST || '192.168.11.104'
const host = process.env.HOST || '172.20.10.3'
const port = process.env.PORT || 3000

let config = require('../nuxt.config.js')
config.dev = !(app.env === 'production')

async function start() {
  const nuxt = new Nuxt(config)

  if (config.dev) {
    const builder = new Builder(nuxt)
    await builder.build()
  }

  await connect()
  initSchemas()

  app
    .use(router.routes())
```

```
    .use(router.allowedMethods())

  app.use(ctx => {
    ctx.status = 200

    return new Promise((resolve, reject) => {
      ctx.res.on('close', resolve)
      ctx.res.on('finish', resolve)
      nuxt.render(ctx.req, ctx.res, promise => {
        promise.then(resolve).catch(reject)
      })
    })
  })

  app.listen(port, host)
  consola.ready({
    message: `Server listening on http://${host}:${port}`,
    badge: true
  })
}

start()
```

需要注意的是 host 要改成实际环境的 IP 地址，并且与前面介绍的 Android 端发起的 URL 地址相匹配。

以上的代码初始化了 Koa、mongoose，调用了 initSchemas 方法对数据库进行初始化。初始化数据库的代码如下所示：

```
const mongoose = require('mongoose')
const Schema = mongoose.Schema

const logSchema = new Schema({
  id: {
    unique: true,
    type: String
  },
  device: String,
  message: String,
```

```
    cause: String,
    trace: [{
      library: String,
      method: String,
      line: Number,
      class: String,
    }],

    meta: {
      createdAt: {
        type: Date,
        default: Date.now()
      },
      updatedAt: {
        type: Date,
        default: Date.now()
      }
    }
})

logSchema.pre('save', function (next) {
  if (this.isNew) {
    this.meta.createdAt = this.meta.updatedAt = Date.now()
  } else {
    this.meta.updatedAt = Date.now()
  }

  next()
})

mongoose.model('FlutterLog', logSchema)
```

这里我们创建了 FluttgerLog 的数据库表名，里面存储的数据库字段是依据 Flutter Client 端创建的。然后，我们通过 Koa2 中间件的接口接收服务端的 post 参数，代码如下所示：

```
router.post('/log', koaBody(), async (ctx) => {
    ctx.body = JSON.stringify(ctx.request.body)
```

```
    const FlutterLog = mongoose.model('FlutterLog')
    var flutterLog = new FlutterLog({
      id: Date.now() + '',
      device: ctx.request.body.device,
      message: ctx.request.body.message,
      cause: ctx.request.body.cause,
      trace: ctx.request.body.trace
    })
    await flutterLog.save()
  }
);
```

调用 save 方法之后会往数据库里面存储一条错误日志数据。最后，我们使用前端的 Vue 框架对获取的异常进行展示。笔者使用 Vue 做了一个简单的界面来展示，代码如下所示：

```
<template>
  <div id="logs-container">
    <ul>
      <li
        v-for="data in logs"
        :key="data.id"
        class="line">
        <p>错误id: {{ data.id }}</p>
        <p>错误机型: {{ data.device }}</p>
        <p>异常名称: {{ data.message }}</p>
        <p>出错位置: {{ data.cause }}</p>
        <ul class="trace-body">
          <li
            v-for="t in data.trace"
            :key="t._id"
            class="trace-line">
            <p>{{ t.library }}</p>
            <p>{{ t.method }}</p>
            <p>{{ t.line }}</p>
            <p>{{ t.class }}</p>
          </li>
        </ul>
```

```
        </li>
      </ul>
    </div>
</template>

<script>
import config from '~/common/config.js'
import axios from 'axios'

export default {
  async asyncData({ query }) {
    const { data } = await axios.get(`${config.baseUrl}/log/find`)
    console.log(data)
    return {
      logs: data.data.logs
    }
  }
}
</script>

<style>
  #logs-container { padding: 10px; font-size: 12px; }
  ul { margin: 0; padding: 0; }
  li { list-style-type: none; }
  .line { margin-bottom: 20px; }
  .trace-body { background-color: #eeeeee; margin: 10px 0; height: 100px; overflow: auto; }
  .trace-line { padding: 3px 10px; }
</style>
```

最后，经过以上步骤的操作，大功告成。我们看一下异常收集的页面，如图 10.8 所示。

回顾一下客户端的实现，如图 10.9 所示。

通过点击界面上的这些按钮，可以在 Android 端发布异常数据并传送给服务端。这样，一个简易的错误日志上传系统就做完了。

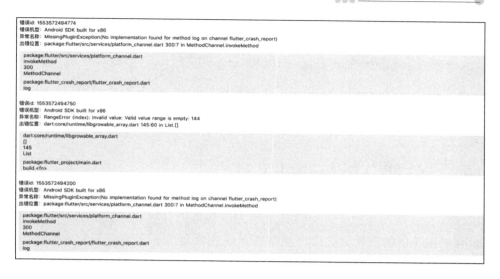

图 10.8

图 10.9

10.3 实战项目源码

本章实战项目对应的源码地址如下所示。

实战一，Flutter 端：chapter10/flutter_practice。

实战一，服务端：chapter10/flutter-practice-server。

实战二，Flutter 端：chapter10/flutter_project。

实战二，Flutter Package 端：chapter10/flutter_crash_report。

实战二，服务端：chapter10/flutter-log-server。

10.4 性能分析与辅助工具

在 App 开发过程中，我们需要对 App 性能做分析，目的是在上线之前解决各种潜在的问题。这个时候就需要使用性能分析工具。我们来看一下 Flutter 为我们提供了哪些性能分析工具。

（1）Flutter Inspector：用于检查 Widget 树，在 App 运行时，它可以通过检查器快速检查布局和定位的问题，如图 10.10 所示。

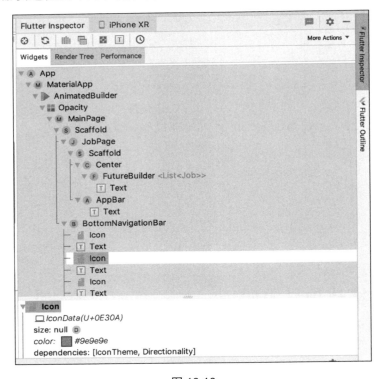

图 10.10

（2）Flutter Outline：可以在一个 Widget 外面包裹另一个 Widget，比如在一个 Text 外面加一层 Row，如图 10.11 所示。

图 10.11

（3）PerformanceOverlay：用于检测 GPU 与 UI 的性能，它通过在代码中添加 showPerformanceOverlay: true 来开启，代码如下所示：

```
@override
Widget build(BuildContext context) {
  return new MaterialApp(
    showPerformanceOverlay: true,
    theme: new ThemeData(
      primaryIconTheme: const IconThemeData(color: Colors.white),
      brightness: Brightness.light,
      primaryColor: Config.GLOBAL_COLOR,
      accentColor: Config.GLOBAL_COLOR,
    ),
    home: SplashPage(),
  );
}
```

效果如图 10.12 所示。

图 10.12

在图 10.12 中，上面是 GPU thread，它表示在 GPU 线程上生成每帧所需要的绘制时间。如果 GPU thread 是红色的，那么可能是绘制的图形过于复杂，或者是执行了过多的 GPU 操作。

下面是 UI thread，表示在 UI 线程上生成每帧所需要的时间。如果 UI thread 是红色的，说明肯定是 Dart 代码里有过度耗时的操作，从而导致了 UI 操作的阻塞。

（4）Dart 分析器：在项目目录下运行"flutter analyze"测试你的代码，分析代码并帮助你发现可能的错误，如图 10.13 所示。

```
info • This class (or a class which this class inherits from) is marked as
       '@immutable', but one or more of its instance fields are not final:
       ScrollImageItem.onPressed • lib/common/widget/scroll_img_item.dart:3:7
       • must_be_immutable
info • This function has a return type of 'Widget', but doesn't end with a
       return statement • lib/main.dart:103:3 • missing_return
info • The class 'Future' was not exported from 'dart:core' until version
       2.1, but this code is required to be able to run on earlier versions •
       lib/widgets/chat_page.dart:15:3 • sdk_version_async_exported_from_core
info • The class 'Future' was not exported from 'dart:core' until version
       2.1, but this code is required to be able to run on earlier versions •
       lib/widgets/company/company_detail_page.dart:32:
       3 •
       sdk_version_async_exported_from_core
info • This function has a return type of 'Future<CompanyDetail>', but
       doesn't end with a return statement •
       lib/widgets/company/company_detail_page.dart:32:3 • missing_return
info • Avoid using braces in interpolation when not needed •
       lib/widgets/company/company_detail_page.dart:389:67 •
       unnecessary_brace_in_string_interps
```

图 10.13

（5）Observatory：通过"flutter run"方式运行之后，在控制台会生成一段 URL，然后把这段 URL 在浏览器中打开，就可以看到图 10.14 所示的信息。

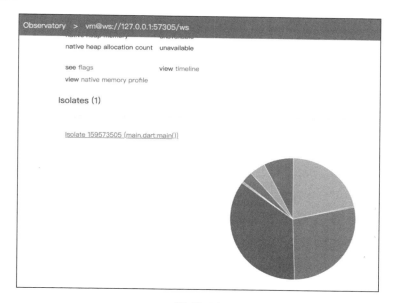

图 10.14

（6）debugDumpRenderTree：尝试调试布局问题，在 Widget 层的树不够详细的情况下，可以通过调用 debugDumpRenderTree 转储渲染树，如图 10.15 所示。

```
flutter: RenderView#f998a
flutter:  │ debug mode enabled - ios
flutter:  │ window size: Size(828.0, 1792.0) (in physical pixels)
flutter:  │ device pixel ratio: 2.0 (physical pixels per logical pixel)
flutter:  │ configuration: Size(414.0, 896.0) at 2.0x (in logical pixels)
flutter:  │ semantics enabled
flutter:  │
flutter:  └─child: RenderSemanticsAnnotations#9a65e
flutter:     │ creator: Semantics ← Localizations ← MediaQuery ←
flutter:     │   WidgetsApp-[GlobalObjectKey _MaterialAppState#0b360] ←
flutter:     │   IconTheme ← _InheritedTheme ← Theme ← AnimatedTheme ←
flutter:     │   ScrollConfiguration ← MaterialApp ← MyApp ← [root]
flutter:     │ parentData: <none>
flutter:     │ constraints: BoxConstraints(w=414.0, h=896.0)
flutter:     │ semantics node: SemanticsNode#1
flutter:     │ size: Size(414.0, 896.0)
flutter:     │
flutter:     └─child: RenderCustomPaint#6eb41
flutter:        │ creator: CustomPaint ← Banner ← CheckedModeBanner ← Title ←
flutter:        │   Directionality ← _LocalizationsScope-[GlobalKey#f8ad4] ←
flutter:        │   Semantics ← Localizations ← MediaQuery ←
flutter:        │   WidgetsApp-[GlobalObjectKey _MaterialAppState#0b360] ←
flutter:        │   IconTheme ← _InheritedTheme ← ←
flutter:        │ parentData: <none> (can use size)
flutter:        │ constraints: BoxConstraints(w=414.0, h=896.0)
flutter:        │ size: Size(414.0, 896.0)
flutter:        │
flutter:        └─child: RenderPointerListener#e34f4
flutter:           │ creator: Listener ← Navigator-[GlobalObjectKey<NavigatorState>
flutter:           │   _WidgetsAppState#52ce0] ← DefaultTextStyle ← CustomPaint ←
flutter:           │   Banner ← CheckedModeBanner ← Title ← Directionality ←
flutter:           │   _LocalizationsScope-[GlobalKey#f8ad4] ← Semantics ←
flutter:           │   Localizations ← MediaQuery ← ←
flutter:           │ parentData: <none> (can use size)
flutter:           │ constraints: BoxConstraints(w=414.0, h=896.0)
flutter:           │ size: Size(414.0, 896.0)
flutter:           │ behavior: deferToChild
flutter:           │ listeners: down, up, cancel
flutter:           │
flutter:           └─child: RenderAbsorbPointer#9603a
flutter:              │ creator: AbsorbPointer ← Listener ←
flutter:              │   Navigator-[GlobalObjectKey<NavigatorState>
flutter:              │   _WidgetsAppState#52ce0] ← DefaultTextStyle ← CustomPaint ←
flutter:              │   Banner ← CheckedModeBanner ← Title ← Directionality ←
flutter:              │   _LocalizationsScope-[GlobalKey#f8ad4] ← Semantics ←
```

图 10.15

其他的命令有 debugDumpApp、debugDumpLayerTree 等。

衡量应用启动时间，可以输入命令 "flutter run—trace-startup—profile"，会保存为 start_up_info.json 文件。

本章小结

本书的内容已经全部讲完了，但是，本书的内容还是有局限性的，你需要在岗位中不断地通过项目实践去提升自己，才能不断进步。Flutter 是一门很新的技术，笔者也在不断地探索中，让我们一起努力学习吧！